OSI Reference Model
for Telecommunications

OSI Reference Model for Telecommunications

Debbra Wetteroth

McGraw-Hill

New York • Chicago • San Francisco • Lisbon
London • Madrid • Mexico City • Milan • New Delhi
San Juan • Seoul • Singapore • Sydney • Toronto

McGraw-Hill

A Division of The McGraw·Hill Companies

Copyright © 2002 by McGraw-Hill Companies, Inc. All rights reserved. Printed in the United States of America. Except as permitted under the United States Copyright Act of 1976, no part of this publication may be reproduced or distributed in any form or by any means, or stored in a data base or retrieval system, without the prior written permission of the publisher.

1 2 3 4 5 6 7 8 9 0   AGM/AGM   0 9 8 7 6 5 4 3 2 1

ISBN 0-07-138041-8

The sponsoring editor for this book was Marjorie Spencer, the editing supervisor was David E. Fogarty, and the production supervisor was Pamela A. Pelton. It was set in Vendome by Patricia Wallenburg.

Printed and bound by Quebocor/Martinsburg.

McGraw-Hill books are available at special quantity discounts to use as premiums and sales promotions, or for use in corporate training programs. For more information, please write to the Director of Special Sales, Professional Publishing, McGraw-Hill, Two Penn Plaza, New York, NY 10121-2298. Or contact your local bookstore.

 This book is printed on recycled, acid-free paper containing a minimum of 50 percent recycled, de-inked fiber.

I dedicate this book to my husband Wayne and my children Crystal and Travis. Their love, support, patience, and encouragement were an essential element to fulfill my book-publishing dream.

CONTENTS

Contents

Contents

Contents

ACKNOWLEDGMENTS

There are many people I would like to thank for making this book a reality.

Wayne for being my devoted husband. The understanding and support you gave me was essential.

Crystal for being my loving daughter. Thank you for the patience you gave me while sharing our quality time with the time I needed to write this book.

Travis for being my loving son. Thank you for understanding when I needed the computer and you could not play your video games.

My mother, Sharon Feldmeier who is always there to listen and offering advice to rebuild my strength.

My father, Carl Feldmeier who encouraged me to pursue my dreams to write this book.

Other Essential Acknowledgments:

Marjorie Spencer, my publisher from McGraw-Hill who made the decision to pursue publishing this book.

Carole McClendon, my agent from Waterside Productions, Incorporated, for establishing the relationship between McGraw-Hill and myself.

Overview

Since divestiture of AT&T in 1982, there has been a growing demand in technology sectors for working knowledge of the telephone industry. At the same time, the industry has evolved tremendously. Changes affect both how telephone systems set up and how they operate under the hood. For an individual first entering the field, it can be overwhelming.

For people in the trenches, *OSI Reference Model for Telecommunications* sets itself apart by describing a complete telecommunications network from the ground up. As in business, you may find it advantageous to start on the ground floor and work your way up. This book will be your companion in that learning process.

To make our network walk through understandable, this book makes valuable use of the Open Systems Interconnection (OSI) Reference Model. By using of this tool, telecommunications network components can be broken down into familiar categories. Each of these working components is defined in relation to the powerful organizing structure of the OSI Reference Model. We will refer to our telecommunications model as the "OSI telecommunications reference model."*

OSI Reference Model in Brief

The International Standards Organization (ISO) produced the Open Systems Interconnection reference model in 1974. Their effort was to standardize network architecture and encourage vendors to develop network equipment that would avoid proprietary design.

The Reference Model comprises seven layers. The higher ones—layers 4 through 7—pertain exclusively to end-to-end functions such as user application, messaging assurance, session establishment, user services, and the user interface. For telecommunications, the "interface" layers (layers 1 through 3) are those that matter. These are the Physical Layer, the Data Link Layer, and the Network Layer. In this book I've tried to shed some light on the OSI Reference Model by further segmenting these first three layers into the hardware, protocols, and topology requirements for each. This book will last you for many years precisely because the initial layers form the stable foundation of any telecommunications network, no matter how changeable the technology.

---

*This book was written for telecommunications professionals; expect to find references to data communication where it relates to telecom applications.

What does it really mean to call layers 1 through 3 "stable"? If you compare a network of today with one built a decade ago, you will find the working components to be the same or analogous in function. Moreover, the initial three layers are where you will configure and troubleshoot your telephone network. By contrast, there's no very useful comparison between today's network and 1990's from the perspective of layers 4 through 7. To meet today's market demands for end-to-end function requires higher bandwidth and faster speeds. Later chapters in this book will cover the Transport Layer technology most relevant to telecom service offerings, and offer high-level descriptions of the most popular transport protocols utilized today.

Table 1.1 compares the OSI Reference Model with what we're calling the OSI Telecommunications Reference Model, so that you can anticipate how the two will map.

TABLE 1.1

OSI Reference Model versus OSI Telecommunications Reference Model

Levels	OSI Reference Model	Telecommunications Model
Level 7	Application	
Level 6	Presentation	
Level 5	Session	
Level 4	Transport	End-to-End Protocols
Level 3	Network	Network
Level 2.5		Multilink
Level 2	Data Link	Data Link
Level 1.5		Logical Link
Level 1	Physical	Physical

Layer Descriptions

- **Layer 1.** The Physical Layer specifies characteristics of the physical data transfer medium for network communications and is also responsible for monitoring data error rates.

- **Layer 1.5.** The Logical Link sublayer defines a service access point and frame format for inter-station exchanges.

- **Layer 2.** The Data Link Layer is responsible for error-free communication between two adjacent nodes. If an error is detected, the Data Link Layer requests a retransmission of the data.

- **Layer 2.5.** The Multilink sublayer provides an interface between the logical links and the physical medium for specific topologies and access control schemes. Logical links enable two communicating Data Link Layers on separate hosts to have common guidelines for flow control, error handling, and data retransmission requests.

- **Layer 3.** The Network Layer is responsible for routing and flow control functions.

- **Layer 4.** The Transport Layer comprises end-to-end protocol functions that increase bandwidth and data-rate speeds.

Telecommunications Standards

Without standards for integration between various carriers and network designs, telecommunications is an empty promise. By definition, telecommunications facilities operate beyond local scope. They access bandwidth over wide-area geographies using the services of carriers such as regional operating companies (RBOCs). Armed with these services, the art of telecommunications provides full-time and part-time connectivity and allows access over various interfaces operating at different speeds.

Telecommunication network design and topology standards are managed by a number of recognized authorities who have worked, nationally and internationally, to assist vendors and developers with the challenge of producing generic network components. The most influential standards organizations for telecommunications include the following agencies:

- International Organization for Standardization (ISO)

  — Based in Geneva, Switzerland

  — Along with ANSI, developed the Open Systems Interconnect (OSI) Reference Model in 1974

- Electronic Industries Association (EIA)

  — Publishes "Recommended Standards" (RS) for physical devices and their means of interfacing

- Institute of Electrical and Electronics Engineers (IEEE)

  — Developed LAN standards for telecommunications networks

  — Set standards for integrated voice and data networks

- International Telecommunication Union-Telecommunication Standardization Sector (ITU-T), formerly known as the Consultative Committee for International Telegraph and Telephone (CCITT)

  — Based in Geneva, Switzerland

  — Modem standards

  — X.400 standard and emerging X.500 recommendations for receiving gateways for translating email files

  — Digital telephone standards

- Electronic Industries Association/Telecommunications Industry Association (EIA/TIA)

  — Joint effort between the Telecommunications Industry Association (TIA) and the Electronic Industries Association (EIA)

  — Developed with the intent of identifying minimum requirements that would support multiproduct and multivendor environments

  — Addresses six elements of cabling specifications and requirements that can affect the type of cable used in a LAN: horizontal cabling, telecommunications closets, equipment rooms, backbone cabling, and entrance facilities

  — Cable standards include EIA/TIA-568A and EIA/TIA-568B categories Three, Four, and Five cable

  — Allow for planning and installation of LAN systems without knowledge of the specific equipment that is to be installed

- American National Standards Institute (ANSI)

  — The United States' representative to the International Standards Organization

  — Along with ISO, developed the Open Systems Interconnect (OSI) Reference Model in 1974

  — Encourages vendors to develop network equipment that doesn't rely on a proprietary design.

Numbering Systems and Character Code Sets

Numbering systems provide a systematic method to read, monitor, analyze, and configure telecommunications transmissions. Data can be transmitted and received in many ways. Letters and decimal numbers are the most convenient for humans, but computers do not easily decipher this information, because computer systems are designed to interpret binary digits only.

A *bit*, which is the abbreviation for binary digit, is used to represent characters, describe measurements of telecommunications data, and encode control information in data fields. Each bit indicates one of two possible data transmission values that are represented by the binary notation characters 0 and 1. The use of 0s and 1s parallels two discrete states of on and off, high and low, or mark and space in electronic circuits. To decipher this information, a method for combining bits to represent numerical values, control characters, and alphabetical quantities must be encoded.

The choice of *bit* or *byte code* length depends entirely on the system design and customer equipment. Two common coding examples (discussed later in the chapter) are Extended Binary Coded Decimal Interexchange Code (EBCDIC) and American Standard Code for Information Interchange (ASCII).

This chapter discusses several numbering systems, beginning with the most common methods and continuing in descending order of popularity. To start off, let's look at the four most common numbering systems in use today—binary, octet, decimal, and hexadecimal.

- The binary system is a numeric-based character code set consisting of 0s and 1s and have a unique 4-bit binary.

- The octet system uses digits 0—7; 8 holds a unique value that is referred to as *base 8*.

- The decimal system uses digits 0—9; 10 holds a unique value that is referred to as *base 10*.

- The hexadecimal system is a group of binary values referred to as *base 16*. Only numbers 0—9 are utilized, so values over 10 are represented by alphabetical symbols A, B, C, D, E, and F.

Decimal Numbering

Decimal numbering is the most common numbering system used in the world. Each decimal position of a number string is represented by a power of 10, which contains digits 0—9. Weight of value is determined by placement of the numeric notation. Each of these positions carries a measurement in the power of 10. Negative powers of 10 are determined by separation of digits with a decimal point.

Figure 2.1 illustrates placement of digits in decimal (base 10) form and their value:

FIGURE 2.1
Decimal notation.

Decimal Number	Power of 10 Exponential Method
1	$10^0 = 1$
10	$10^1 = 10 \times 1$
100	$10^2 = 10 \times 10$

Figure 2.2 illustrates an exponential method of a decimal value as a *positive* power of 10. As illustrated here any number raised to the 0 power is equal to 1.

FIGURE 2.2
Exponential decimal notation.

One	1.
Ten	10.
One Hundred	100.
One Tenth	.1
One Hundredth	.01

Figure 2.3 illustrates the exponential method for decimal number having a *negative* power of 10.

FIGURE 2.3
Negative
exponentiation.

Decimal Number	Power of 10 Exponential Method
.1	$10^{-1} = 10 \times -1$
.01	$10^{-2} = 10 \times -10$

Binary Numbering

Using the binary numbering system implies a selection, choice, or condition of data transmission digits that have the possibility of two different values or states. This numeric-based character code set consists of numbers 0—9 and has a unique 4-bit binary. The binary numbering scheme is just as systematic as decimal or base 10.

Binary numbers consist of a base 2 numbering system of 0s and 1s (i.e. 01, 0110, 01101110). An 8-digit character notation of 0s and 1s is most commonly used, and just as in the decimal numbering system, the arrangement of bits in a sequence determines weight. A string of 0s and 1s with a subscript of 2 indicates a binary number (i.e., 11101101_2).

Placement of a binary number is expressed by the power of 2. If a 1 is placed in a sequence, the value of that placement *is* counted. If a 0 is placed, the value of that placement *is not* counted. This method of numbering system is primarily used in an "all or nothing" environment such as circuitry.

Figure 2.4 illustrates the weight of the binary number 110101011_2. Notice how each decimal weight field from right to left increases by a power of 2.

FIGURE 2.4
Binary numbers.

Decimal Weight	256	128	64	32	16	8	4	2	1_{10}	
Binary Number	1	1	0	1	0	1	0	1	1_2	
Bring Down Decimal Weight if Binary Number = 1	256	128	0	32	0	8	0	2	1_{10}	Total = 427_{10}

Figure 2.5 illustrates the power of 2.

FIGURE 2.5
Powers of 2.

Raised to Power of 2	Decimal Weight
$2^0 = 1$	1
$2^1 = 2 \times 1$	2
$2^2 = 2 \times 2$	4
$2^3 = 2 \times 2 \times 2$	8
$2^4 = 2 \times 2 \times 2 \times 2$	16
$2^5 = 2 \times 2 \times 2 \times 2 \times 2$	32
$2^6 = 2 \times 2 \times 2 \times 2 \times 2 \times 2$	64
$2^7 = 2 \times 2 \times 2 \times 2 \times 2 \times 2 \times 2$	128
$2^8 = 2 \times 2 \times 2 \times 2 \times 2 \times 2 \times 2 \times 2$	256

Figure 2.6 illustrates the conversion of binary (base 2) values to decimal (base 10) form and value.

FIGURE 2.6
Conversion of binary
to decimal values.

Power of 2	2^8	2^7	2^6	2^5	2^4	2^3	2^2	2^1	2^0	
Decimal Weight Conversion	256	128	64	32	16	8	4	2	1_{10}	
Binary Number	1	1	0	1	0	1	0	1	1_2	
Bring Down Decimal Weight if Binary Number = 1	256	128	0	32	0	8	0	2	1_{10}	Total = 427_{10}

The base 2 numbering scheme can be used to represent the same amount of numbers as base 10.

Octet Numbers

Another numeric base term found in communication protocols is the octet. Octet numbering is a base 8 numbering system. The bits within the octet may not be related: Three bits may be used for a transmit sequence number, three other bits for a receive acknowledgment number, one bit to give permission to transmit, and one bit to identify the message type.

The octet numbering scheme is just as systematic as decimal (base 10) and binary (base 2). A character code set containing a subscript eight ($_8$) indicates an octet number (i.e., 7_8). Just as in binary numbering, placement of octal digits determines weight; however, the weight of an octet is to the power of 8.

Figure 2.7 illustrates an octet (power of 8) to decimal (power of 10) conversion. Figure 2.8 cross-references octet and decimal number values.

FIGURE 2.7
Octet to decimal conversion.

Raised to Power of 16	Decimal Weight
$1^0 = 1$	1
$8^1 = 8 \times 1$	8
$8^2 = 8 \times 8$	64
$8^3 = 8 \times 8 \times 8$	512
$8^4 = 8 \times 8 \times 8 \times 8$	4,096
$8^5 = 8 \times 8 \times 8 \times 8 \times 8$	32,768
$8^6 = 8 \times 8 \times 8 \times 8 \times 8 \times 8$	262,144

Octet	0	1	2	3	4	5	6	7	10	11	12	13	14	15	16	17
Decimal	0	1	2	3	4	5	6	7	8	9	10	11	12	13	14	15

FIGURE 2.8   Cross-reference of octal (top row) to decimal numbers.

Hexadecimal Numbers

The hexadecimal system is a base 16 numbering system. Because communication system components, such as analyzers that monitor the telecommunication performance of a circuit and tools that analyze files on a floppy disc, usually work with eight-bit groups, hexadecimal notation is a perfect fit to represent these values. Each hexadecimal value corresponds to a 4-bit binary numeric-based character code set. Decimal numbers 0 to 15 are represented using a single symbol—a normal numeric 0—9 and an alpha character, A—F. Two 4-bit groups comprise one *byte* and are expressed as a two-character hexadecimal value. When grouping binary digits into sets of four bits each, hexadecimal is an easily readable shorthand for binary.

The hexadecimal numbering scheme is just as systematic as decimal (base 10), binary (base 2), and octet (base 8). A character code set containing subscript sixteen ($_{16}$) indicates a hexadecimal number (i.e., $B5_{16}$). Just as in decimal and binary, the placement of hexadecimal digits determines weight; however, the weight of a hexadecimal is to the power of 16.

Figure 2.9 illustrates a hexadecimal (power of 16) to decimal (power of 10) conversion.

FIGURE 2.9
Hexadecimal to
decimal conversion.

Raised to Power of 16	Decimal Weight
$16^0 = 1$	1
$16^1 = 16 \times 1$	16
$16^2 = 16 \times 16$	256
$16^3 = 16 \times 16 \times 16$	4,096
$16^4 = 16 \times 16 \times 16 \times 16$	65,536
$16^5 = 16 \times 16 \times 16 \times 16 \times 16$	1,048,576
$16^6 = 16 \times 16 \times 16 \times 16 \times 16 \times 16$	16,777,216

Figure 2.10 cross-references hexadecimal and decimal values.

Hexadecimal	0	1	2	3	4	5	6	7	8	9	A	B	C	D	E	F
Decimal	0	1	2	3	4	5	6	7	8	9	10	11	12	13	14	15

FIGURE 2.10   Cross-reference of hexadecimal (top row) and decimal numbers.

Figure 2.11 compares values in the four numbering systems discussed.

FIGURE 2.11 Comparison of binary, octet, decimal, and hexadecimal values.

BASE STRUCTURE			
Binary = base 2	Octet = base 8	Decimal = base 10	Hexadecimal = base 16
0000	0	0	0
0001	1	1	1
0010	2	2	2
0011	3	3	3
0100	4	4	4
0101	5	5	5
0110	6	6	6
0111	7	7	7
1000	10	8	8
1001	11	9	9
1010	12	10	A
1011	13	11	B
1100	14	12	C
1101	15	13	D
1110	16	14	E
1111	17	15	F

Number System Conversions

Decimal to Hexadecimal Conversion

Convert decimal 9510 to a hexadecimal value:

* Divide the decimal number by the hexadecimal base, which is 16 (95 divided by 16 equals 5 with a remainder of 15)

* Refer to Figure 2.10 to acquire the decimal-hexadecimal cross-reference value:

  — 5 is a valid hexadecimal

  — 15 is converted to F

  — Answer is: $5F_{16}$

Binary to Hexadecimal Conversion

To convert a binary number to a hexadecimal number, separate the binary number into groups of four digits starting at the extreme right. Figure 2.12 illustrates this step using the binary example 11011011_2.

FIGURE 2.12
Binary to hexadecimal conversion, step 1.

Binary Number	1	1	0	1_2	1	0	1	1_2

The *decimal* weights of each bit must be first assigned to each group. (Remember that we are converting binary to decimal, so the decimal weight is in power of 2.) Figure 2.13 illustrates this step.

FIGURE 2.13
Binary to hexadecimal conversion, step 2.

Decimal Weight	8	4	2	1_{10}	8	4	2	1_{10}
Binary Number	1	1	0	1_2	1	0	1	1_2

Figure 2.14 determines the total of each group by adding the weight of all 1s.

Decimal Weight	8	4	2	1 $_{10}$		8	4	2	1 $_{10}$	
Binary Number	1	1	0	1 $_{2}$		1	0	1	1 $_{2}$	
Bring Down if Binary Number is = 1	8	4		1 $_{10}$	13 = D	8		2	1 $_{10}$	11 = B Hexadecimal = DB

FIGURE 2.14   Binary to hexadecimal conversion, step 3.

Therefore, binary $11011011_2 = DB_{16}$ in hexadecimal.

Hexadecimal to Binary Conversion

Converting hexadecimal to binary is the opposite of binary to hexadecimal. Remember, hexadecimal and binary conversions always have to go through decimal conversions first. Figure 2.15 illustrates a decimal weight divided into two separate 4-bit groups.

FIGURE 2.15
Hexadecimal to
binary conversion,
step 1.

Decimal Weight in Power of 2 (Binary)	8	4	2	1 $_{10}$	8	4	2	1 $_{10}$

Figure 2.16 uses the hexadecimal value to find the combination of decimal values to sum up to the hexadecimal value.

Decimal Weight in Power of 2 (Binary)	8	4	2	1 $_{10}$		8	4	2	1 $_{10}$	
Decimal Weights = Hexadecimal Value	8	4		1 $_{10}$	D = 13	8		2	1 $_{10}$	B = 11

FIGURE 2.16   Hexadecimal to binary conversion, step 2.

In Figure 2.17, we place a binary 1 in the field where a decimal value is counted in the sum of the hexadecimal and a 0 in the field where it is not.

Decimal Weight in Power of 2 (Binary)	8	4	2	1 $_{10}$		8	4	2	1 $_{10}$	
Decimal Weights = Hexadecimal Value	8	4		1 $_{10}$	D = 13	8		2	1	B = 11
Binary Values	1	1	0	1 $_{2}$		1	0	1	1 $_{2}$	

FIGURE 2.17   Hexadecimal to binary conversion, step 3.

Reconnect the two 4-bit groups of binary values and you now have hexadecimal DB^{16} converted to 11011011_2.

Units of Measurement

A *bit* is a single unit of measurement. When multiple bits are formed into a group to represent a number, letter, or character, it is called a *byte*. The most common size byte is a grouping of 8 bits. When bytes are grouped in larger blocks, the measurement is used *kilobyte*, which literally means "a thousand bytes." (When referring to computer systems, however, a kilobyte is really 1,024 bytes, because the power of 2 is used.)

An octet is a group of 8 bits. Each bits within this 8-bit grouping may be used for different purposes—1 bit for receipt of packet, 3 bits for the acknowledgment, 3 bits for synchronization, and 1 bit for packet type. All 8 bits together are called an *octet.*

Character Code Sets

It is important to understand numbering systems and units of measurement, but these concepts hold little value unless we can use them to convey information. As we have noted, a bit can only have two states, 0 or 1. To form a character, the combination of several bits is required. There are only four combinations possible when using two bits—00, 01, 10, and 11. If we use three bits, then eight combinations are possible. As we add additional bits, computation is based on 2 to the power of the number of bits.

Figure 2.18 illustrates the different values arrived at each time a bit is added.

FIGURE 2.18
Adding bits to a
binary number.

2_2	2_1	2_0	Decimal
0	0	0	0
0	0	1	1
0	1	0	2
0	1	1	3
1	0	0	4
1	0	1	5
1	1	0	6
1	1	1	7

Within an arrangement of bits, these characters must be represented in order to convey information:

- 10 numeric symbols

- 26 letters (upper- and lowercase, so actually 52 characters)

- Special characters (+, −, %, $, etc.)

- Punctuation characters (periods, commas, question marks, etc.)

- Control characters (LineFeed, CarriageReturn, etc.)

Three coding schemes are commonly used to convey data:

- ASCII

- Baudot

- EBCDIC

American Standard Code for Information Interchange (ASCII)

The American Standard Code for Information Interchange (ASCII), is the most commonly used code system in the United States and is also widely used outside the United States. There are a number of standardized versions with different names, but basically these refer to the same code.

The International Consultative Committee on Telegraphy and Telephone (CCIATT) has a version called International Alphabet No. 5 (IA5), and the International Standards Organization (ISO) has produced a standard called ISO Seven-Bit Coded Character Set for Information-Processing Interchange. There are national options available within the code so that a region can elect to use special characters that are unique to an area or language.

Because computers are capable of synchronization using 8 bits, ASCII is the cleanest way to represent this information. ASCII is an 8-level or 8-bit code consisting of seven information bits plus a *parity bit*. This eighth or parity bit is used for error detection. Each character in the seven information bits is represented by a unique 7-bit pattern; thus if we calculate 7 bits to the power of 2, we have 128 possible bit combinations.

Characters are coded and read from left to right. Bit positions are 7 to 1.

- Example: $b_7\ b_6\ b_5\ b_4\ b_3\ b_2\ b_1$

- Bit 7 is considered the *Most Significant Bit (MSB)*

- Bit 1 is the *Least Significant Bit (LSB)*

Table 2.1 illustrates the ASCII Character Set.

TABLE 2.1 ACSII Character Code Conversion Chart

Binary	Octet	Decimal	Hex	Character Code	Binary	Octet	Decimal	Hex	Character Code
0000000_2	0_8	0_{10}	00_{16}	NUL	1000000_2	100_8	64_{10}	40_{16}	@
0000001_2	1_8	1_{10}	01_{16}	(TC$_1$)SOH	1000001_2	101_8	65_{10}	41_{16}	A
0000010_2	2_8	2_{10}	02_{16}	(TC$_2$)STX	1000010_2	102_8	66_{10}	42_{16}	B
0000011_2	3_8	3_{10}	03_{16}	(TC$_3$)ETX	1000011_2	103_8	67_{10}	43_{16}	C
0000100_2	4_8	4_{10}	04_{16}	(TC$_4$)EOT	1000100_2	104_8	68_{10}	44_{16}	D
0000101_2	5_8	5_{10}	05_{16}	(TC$_5$)ENQ	1000101_2	105_8	69_{10}	45_{16}	E
0000110_2	6_8	6_{10}	06_{16}	(TC$_6$)ACK	1000110_2	106_8	70_{10}	46_{16}	F
0000111_2	7_8	7_{10}	07_{16}	BEL	1000111_2	107_8	71_{10}	47_{16}	G
0001000_2	10_8	8_{10}	08_{16}	BS	1001000_2	110_8	72_{10}	48_{16}	H
0001001_2	11_8	9_{10}	09_{16}	(FE$_1$)HT	1001001_2	111_8	73_{10}	49_{16}	I
0001010_2	12_8	10_{10}	$0A_{16}$	(FE$_2$)LF	1001010_2	112_8	74_{10}	$4A_{16}$	J
0001011_2	13_8	11_{10}	$0B_{16}$	(FE$_3$)VT	1001011_2	113_8	75_{10}	$4B_{16}$	K
0001100_2	14_8	12_{10}	$0C_{16}$	(FE$_4$)FF	1001100_2	114_8	76_{10}	$4C_{16}$	L
0001101_2	15_8	13_{10}	$0D_{16}$	(FE$_5$)CR	1001101_2	115_8	77_{10}	$4D_{16}$	M
0001110_2	16_8	14_{10}	$0E_{16}$	SO	1001110_2	116_8	78_{10}	$4E_{16}$	N
0001111_2	17_8	15_{10}	$0F_{16}$	SI	1001111_2	117_8	79_{10}	$4F_{16}$	O
0010000_2	20_8	16_{10}	10_{16}	(TC$_7$)DLE	1010000_2	120_8	80_{10}	50_{16}	P
0010001_2	21_8	17_{10}	11_{16}	DC$_1$	1010001_2	121_8	81_{10}	51_{16}	Q
0010010_2	22_8	18_{10}	12_{16}	DC$_2$	10100102	122_8	82_{10}	52_{16}	R
0010011_2	23_8	19_{10}	13_{16}	DC$_3$	10100112	123_8	83_{10}	53_{16}	S
0010100_2	24_8	2010	14_{16}	DC$_4$	10101002	124_8	84_{10}	54_{16}	T

continued on next page

TABLE 2.1 ACSII Character Code Conversion Chart (Continued)

Binary	Octet	Decimal	Hex	Character Code	Binary	Octet	Decimal	Hex	Character Code
0010101_2	25_8	21_{10}	15_{16}	$(TC_8)NAK$	1010101_2	125_8	85_{10}	55_{16}	U
0010110_2	26_8	22_{10}	16_{16}	$(TC_8)SYN$	1010110_2	126_8	86_{10}	56_{16}	V
0010111_2	27_8	23_{10}	17_{16}	$(TC_9)ETB$	1010111_2	127_8	87_{10}	57_{16}	W
0011000_2	30_8	24_{10}	18_{16}	CAN	1011000_2	130_8	88_{10}	58_{16}	X
0011001_2	31_8	25_{10}	19_{16}	EM	1011001_2	131_8	89_{10}	59_{16}	Y
0011010_2	32_8	26_{10}	$1A_{16}$	SUB	1011010_2	132_8	90_{10}	$5A_{16}$	Z
0011011_2	33_8	27_{10}	$1B_{16}$	ESC	1011011_2	133_8	91_{10}	$5B_{16}$	[
0011100_2	34_8	28_{10}	$1C_{16}$	$(IS_4)FS$	1011100_2	134_8	92_{10}	$5C_{16}$	\
0011101_2	35_8	29_{10}	$1D_{16}$	$(IS_3)GS$	1011101_2	135_8	93_{10}	$5D_{16}$	]
0011110_2	36_8	30_{10}	$1E_{16}$	$(IS_2)RS$	1011110_2	136_8	94_{10}	$5E_{16}$	^
0011111_2	37_8	31_{10}	$1F_{16}$	$(IS_1)US$	1011111_2	137_8	95_{10}	$5F_{16}$	_
0100000_2	40_8	32_{10}	20_{16}	SP	1100000_2	140_8	96_{10}	60_{16}	`
0100001_2	41_8	33_{10}	21_{16}	!	1100001_2	141_8	97_{10}	61_{16}	a
0100010_2	42_8	34_{10}	22_{16}	"	1100010_2	142_8	98_{10}	62_{16}	b
0100011_2	43_8	35_{10}	23_{16}	#	1100011_2	143_8	99_{10}	63_{16}	c
0100100_2	44_8	36_{10}	24_{16}	$	1100100_2	144_8	100_{10}	64_{16}	d
0100101_2	45_8	37_{10}	25_{16}	%	1100101_2	145_8	101_{10}	65_{16}	e
0100110_2	46_8	38_{10}	26_{16}	&	1100110_2	146_8	102_{10}	66_{16}	f
0100111_2	47_8	39_{10}	27_{16}	'	1100111_2	147_8	103_{10}	67_{16}	g
0101000_2	50_8	40_{10}	28_{16}	(	1101000_2	150_8	104_{10}	68_{16}	h
0101001_2	51_8	41_{10}	29_{16}	)	1101001_2	151_8	105_{10}	69_{16}	i
0101010_2	52_8	42_{10}	$2A_{16}$	*	1101010_2	152_8	106_{10}	$6A_{16}$	j
0101001_2	53_8	43_{10}	$2B_{16}$	+	1101011_2	153_8	107_{10}	$6B_{16}$	k

continued on next page

TABLE 2.1 *ACSII Character Code Conversion Chart (Continued)*

Binary	Octet	Decimal	Hex	Character Code	Binary	Octet	Decimal	Hex	Character Code
0101100_2	54_8	44_{10}	$2C_{16}$	,	1101100_2	154_8	108_{10}	$6C_{16}$	l
0101101_2	55_8	45_{10}	$2D_{16}$	-	1101101_2	155_8	109_{10}	$6D_{16}$	m
0101110_2	56_8	46_{10}	$2E_{16}$	.	1101110_2	156_8	110_{10}	$6E_{16}$	n
0101111_2	57_8	47_{10}	$2F_{16}$	/	1101111_2	157_8	111_{10}	$6F_{16}$	o
0110000_2	60_8	48_{10}	30_{16}	0	1110000_2	160_8	112_{10}	70_{16}	p
0110001_2	61_8	49_{10}	31_{16}	1	1110001_2	161_8	113_{10}	71_{16}	q
0110010_2	62_8	50_{10}	32_{16}	2	1110010_2	162_8	114_{10}	72_{16}	r
0110011_2	63_8	51_{10}	33_{16}	3	1110011_2	163_8	115_{10}	73_{16}	s
0110100_2	64_8	52_{10}	34_{16}	4	1110100_2	164_8	1_{1610}	74_{16}	t
0110101_2	65_8	53_{10}	35_{16}	5	1110101_2	165_8	117_{10}	75_{16}	u
0110110_2	66_8	54_{10}	36_{16}	6	1110110_2	166_8	118_{10}	76_{16}	v
0110111_2	67_8	55_{10}	37_{16}	7	1110111_2	167_8	119_{10}	77_{16}	w
0111000_2	70_8	56_{10}	38_{16}	8	1111000_2	170_8	120_{10}	78_{16}	x
0111001_2	71_8	57_{10}	39_{16}	9	1111001_2	171_8	121_{10}	79_{16}	y
0111010_2	72_8	58_{10}	$3A_{16}$	:	1111010_2	172_8	122_{10}	$7A_{16}$	z
0111011_2	73_8	59_{10}	$3B_{16}$	;	1111011_2	173_8	123_{10}	$7B_{16}$	{
0111100_2	74_8	60_{10}	$3C_{16}$	<	1111100_2	174_8	124_{10}	$7C_{16}$		
0111101_2	75_8	61_{10}	$3D_{16}$	=	1111101_2	175_8	125_{10}	$7D_{16}$	}
0111110_2	76_8	62_{10}	$3E_{16}$	>	1111110_2	176_8	126_{10}	$7E_{16}$	~
0111111_2	77_8	63_{10}	$3F_{16}$	?	1111111_2	177_8	127_{10}	$7F_{16}$	DEL

Great table, right? But what do all these control characters mean? I'll explain in the next six sections.

GENERIC CONTROL CHARACTERS.   ASCII code has four generic classes of control characters as well as a number of individual characters:

- Transmission Controls (TC_1—TC_{10})
- Format Effectors (FE_0—FE_5)
- Device Controls (DC_1—DC_4)
- Information Separators (IS_1—IS_4)

TRANSMISSION CONTROLS (TC_1–TC_{10}).   Transmission control characters are used to:

- Frame a message associated with character-oriented protocols such as binary synchronous communication
- Control the flow of data in a network

Figure 2.19 is a list of transmission control characters and their acronyms.

FIGURE 2.19
ASCII transmission control characters.

TC_1	SOH	Start Of Heading
TC_2	STX	Start Of Text
TC_3	ETX	End Of Text
TC_4	EOT	End Of Transmission
TC_5	ENQ	ENQuiry
TC_6	ACK	ACKnowledge
TC_7	DLE	Data Link Escape
TC_8	NAK	Negative AcKnowledgment
TC_9	SYN	SYNchronous/idle
TC_{10}	ETB	End of Transmission Block

FORMAT EFFECTORS (FE_0–FE_5).   Format effectors control the physical layout of information transmitted to the printed page or terminal screen. Figure 2.20 is a list of format effectors and their acronyms.

FE_0	BS	Back Space
FE_1	HT	Horizontal Tab
FE_2	LF	Line Feed
FE_3	VT	Vertical Tab
FE_4	FF	Form Feed
FE_5	CR	Carriage Return

DEVICE CONTROLS (DC_1–DC_4). Device controls are used primarily to control ancillary devices or special terminal features and for flow control when transmitting data to simple character-oriented asynchronous terminals. These characters halt the flow of data from the host computer to the terminal when a communication fault exists. Figure 2.21 lists the device control characters and their acronyms.

DC_1	XONN	Device Control ON
DC_2	XONN	Device Control ON
DC_3	XOFF	Device Control OFF
DC_4	XOFF	Device Control OFF

INFORMATION SEPARATORS (IS_1–IS_4). Information separators are used to logically delimit data in hierarchical order. Information separators can be combined with format effectors and other characters for the same purpose of logically delimiting data. Figure 2.22 is a list of information separator characters and their acronyms.

IS_1	US	UNIT SEPARATOR
IS_2	RS	RECORD SEPARATOR
IS_3	GS	GROUP SEPARATOR
IS_4	FS	FILE SEPARATOR

INDIVIDUAL CONTROL CHARACTERS. The control characters listed in Figure 2.23 are used to represent common keyboard events.

FIGURE 2.23
ACSII individual
control characters.

ESC	ESCAPE
DEL	DELETE
NUL	NULL
BEL	BELL

Extended Binary Coded Decimal Interchange Code (EBCDIC)

Extended Binary Coded Decimal Interchange Code (EBCDIC) was developed by IBM. It is an 8-bit code in which each character is represented by a unique 8-bit pattern. A unique eight-bit pattern, at the power of 2, equals 256 (2^8) different characters that can be represented.

The format of EBCDIC code is similar to that of ASCII. One difference between EBCDIC and ASCII is that ASCII uses a parity bit for error checking; EBCDIC uses other means. Another major difference is that the EBCDIC bit positions are the exact opposite of the ASCII bit positions—when viewing Figure 2.24, remember that EBCDIC characters are coded and read from left to right. Bit positions are 0 to 7; the Most Significant Bit (MSB) is b_0 and the Least Significant Bit is b_7.

Figure 2.24 compares the bit arrangements for ASCII and EBCDIC.

FIGURE 2.24
ASCII (on left) vs.
EBCDIC code.

ASCII	**EBCDIC**
B_7 B_6 B_5 B_4 B_3 B_2 B_1 B_0	B_0 B_1 B_2 B_3 B_4 B_5 B_6 B_7

EBCDIC, BINARY, AND HEXADECIMAL CHART. Table 2.2 is a chart reference illustrating the EBCDIC Character Code Set and its related binary and hexadecimal numbers.

TABLE 2.2 EBCDIC Character Code Set with Binary and Hexadecimal Equivalents

Decimal Number	Binary Number	Hex	Character Code	Decimal Number	Binary Number	Hex	Character Code	Decimal Number	Binary Number	Hex	Character Code
0_{10}	00000000_2	00_{16}	NUL	53_{10}	00110111_2	37_{16}	EOT	106_{10}	10011000_2	98_{16}	q
1_{10}	00000001_2	01_{16}	SOH	54_{10}	00111000_2	38_{16}	SBS	107_{10}	10011001_2	99_{16}	r
2_{10}	00000010_2	02_{16}	STX	55_{10}	00111001_2	39_{16}	IT	108_{10}	10100001_2	$A1_{16}$	~
3_{10}	00000011_2	03_{16}	ETX	56_{10}	00111010_2	$3A_{16}$	RFF	109_{10}	10100010_2	$A2_{16}$	s
4_{10}	00000100_2	04_{16}	SEL	57_{10}	00111011_2	$3B_{16}$	CU3	110_{10}	10100011_2	$A3_{16}$	t
5_{10}	00000101_2	05_{16}	HT	58_{10}	00111100_2	$3C_{16}$	DC4	111_{10}	10100100_2	$A4_{16}$	u
6_{10}	00000110_2	06_{16}	RNL	59_{10}	00111101_2	$3D_{16}$	NAK	112_{10}	10100101_2	$A5_{16}$	v
7_{10}	00000111_2	07_{16}	DEL	60_{10}	00111111_2	$3F_{16}$	SUB	113_{10}	10100110_2	$A6_{16}$	w
8_{10}	00001000_2	08_{16}	GE	61_{10}	01000000_2	40_{16}	SP	114_{10}	10100111_2	$A7_{16}$	x
9_{10}	00001001_2	09_{16}	SPS	62_{10}	01000001_2	41_{16}	RSP	115_{10}	10101000_2	$A8_{16}$	y
10_{10}	00001010_2	$0A_{16}$	RPT	63_{10}	01001010_2	$4A_{16}$	¢	116_{10}	10101001_2	$A9_{16}$	Z
11_{10}	00001011_2	$0B_{16}$	VT	64_{10}	01001011_2	$4B_{16}$	.	117_{10}	11000000_2	CO_{16}	{
12_{10}	00001100_2	$0C_{16}$	FF	65_{10}	01001100_2	$4C_{16}$	<	118_{10}	11000001_2	$C1_{16}$	A
13_{10}	00001101_2	$0D_{16}$	CR	66_{10}	01001101_2	$4D_{16}$	(	119_{10}	11000010_2	$C2_{16}$	B
14_{10}	00001110_2	$0E_{16}$	SO	67_{10}	01001110_2	$4E_{16}$	+	120_{10}	11000011_2	$C3_{16}$	C
15_{10}	00001111_2	$0F_{16}$	SI	68_{10}	01001111_2	$4F_{16}$	|	121_{10}	11000100_2	$C4_{16}$	D
16_{10}	00010000_2	10_{16}	DLE	69_{10}	01010000_2	50_{16}	&	122_{10}	11000101_2	$C5_{16}$	E
17_{10}	00010001_2	11_{16}	DC_1	70_{10}	01011010_2	$5A_{16}$	!	123_{10}	11000110_2	$C6_{16}$	F
18_{10}	00010010_2	12_{16}	DC_2	71_{10}	01011011_2	$5B_{16}$	$	124_{10}	11000111_2	$C7_{16}$	G

continued on next page

TABLE 2.2 EBCDIC Character Code Set with Binary and Hexadecimal Equivalents (Continued)

Decimal Number	Binary Number	Hex	Character Code	Decimal Number	Binary Number	Hex	Character Code	Decimal Number	Binary Number	Hex	Character Code
19_{10}	00010011_2	13_{16}	DC$_3$	72_{10}	01011100_2	$5C_{16}$	.	125_{10}	11001000_2	$C8_{16}$	H
20_{10}	00010100_2	14_{16}	RES/ENP	73_{10}	01011101_2	$5D_{16}$	)	126_{10}	11001001_2	$C9_{16}$	I
21_{10}	00010101_2	15_{16}	NL	74_{10}	01011110_2	$5E_{16}$	;	127_{10}	11001010_2	CA_{16}	SHY
22_{10}	00010110_2	16_{16}	BS	75_{10}	01011111_2	$5F_{16}$		128_{10}	11010000_2	$D0_{16}$	}
23_{10}	00010111_2	17_{16}	POC	76_{10}	01101001_2	69_{16}	\	129_{10}	11010001_2	$D1_{16}$	J
24_{10}	00011000_2	18_{16}	CAN	77_{10}	01101010_2	$6A_{16}$	|	130_{10}	11010010_2	$D2_{16}$	K
25_{10}	00011001_2	19_{16}	EM	78_{10}	01101011_2	$6B_{16}$	,	131_{10}	11010011_2	$D3_{16}$	L
26_{10}	00011010_2	$1A_{16}$	UBS	79_{10}	01101100_2	$6C_{16}$	%	132_{10}	11010100_2	$D4_{16}$	M
27_{10}	00011011_2	$1B_{16}$	CU1	80_{10}	01101101_2	$6D_{16}$	_	133_{10}	11010101_2	$D5_{16}$	N
28_{10}	00011100_2	$1C_{16}$	IFS	81_{10}	01101110_2	$6E_{16}$	>	134_{10}	11010110_2	$D6_{16}$	O
29_{10}	00011101_2	$1D_{16}$	IGS	82_{10}	01101111_2	$6F_{16}$	?	135_{10}	11010111_2	$D7_{16}$	P
30_{10}	00011110_2	$1E_{16}$	IRS	83_{10}	01111001_2	79_{16}	`	136_{10}	11011000_2	$D8_{16}$	Q
31_{10}	00011111_2	$1F_{16}$	IUB/ITB	84_{10}	01111010_2	$7A_{16}$	:	137_{10}	11011001_2	$D9_{16}$	R
32_{10}	00100000_2	20_{16}	DS	85_{10}	01111011_2	$7B_{16}$	#	138_{10}	11100000_2	$E0_{16}$	\
33_{10}	00100001_2	21_{16}	SOS	86_{10}	01111100_2	$7C_{16}$	@	139_{10}	11100001_2	$E1_{16}$	NSP
34_{10}	00100010_2	22_{16}	FS	87_{10}	01111101_2	$7D_{16}$	'	140_{10}	11100010_2	$E2_{16}$	S
35_{10}	00100011_2	23_{16}	WUS	88_{10}	01111110_2	$7E_{16}$	=	141_{10}	11100011_2	$E3_{16}$	T
36_{10}	00100100_2	24_{16}	BYP/INP	89_{10}	01111111_2	$7F_{16}$	"	142_{10}	11100100_2	$E4_{16}$	U
37_{10}	00100101_2	25_{16}	LF	90_{10}	10000001_2	81_{16}	a	143_{10}	11100101_2	$E5_{16}$	V

continued on next page

TABLE 2.2 EBCDIC Character Code Set with Binary and Hexadecimal Equivalents (continued)

Decimal Number	Binary Number	Hex	Character Code	Decimal Number	Binary Number	Hex	Character Code	Decimal Number	Binary Number	Hex	Character Code
38_{10}	00100110_2	26_{16}	ETB	91_{10}	10000010_2	82_{16}	b	144_{10}	11100110_2	$E6_{16}$	W
39_{10}	00100111_2	27_{16}	ESC	92_{10}	10000011_2	83_{16}	c	145_{10}	11100111_2	$E7_{16}$	X
40_{10}	00101000_2	28_{16}	SA	93_{10}	10000100_2	84_{16}	d	146_{10}	11101000_2	$E8_{16}$	Y
41_{10}	00101001_2	29_{16}	SFE	94_{10}	10000101_2	85_{16}	e	147_{10}	11101001_2	$E9_{16}$	Z
42_{10}	00101010_2	$2A_{16}$	SM/SW	95_{10}	10000110_2	86_{16}	f	148_{10}	11110000_2	$F0_{16}$	0
43_{10}	00101001_2	$2B_{16}$	CSP	96_{10}	10000111_2	87_{16}	g	149_{10}	11110001_2	$F1_{16}$	1
44_{10}	00101100_2	$2C_{16}$	MFA	97_{10}	10001000_2	88_{16}	h	150_{10}	11110010_2	$F2_{16}$	2
45_{10}	00101101_2	$2D_{16}$	ENQ	98_{10}	10001001_2	89_{16}	i	151_{10}	11110011_2	$F3_{16}$	3
46_{10}	00101110_2	$2E_{16}$	ACK	99_{10}	10010001_2	91_{16}	j	152_{10}	11110100_2	$F4_{16}$	4
47_{10}	00101111_2	$2F_{16}$	BEL	100_{10}	10010010_2	92_{16}	k	153_{10}	11110101_2	$F5_{16}$	5
48_{10}	00110010_2	32_{16}	SYN	101_{10}	10010011_2	93_{16}	l	154_{10}	11110110_2	$F6_{16}$	6
49_{10}	00110011_2	33_{16}	IR	102_{10}	10010100_2	94_{16}	m	155_{10}	11110111_2	$F7_{16}$	7
50_{10}	00110100_2	34_{16}	PP	103_{10}	10010101_2	95_{16}	n	156_{10}	11111000_2	$F8_{16}$	8
51_{10}	00110101_2	35_{16}	TRN	104_{10}	10010110_2	96_{16}	o	157_{10}	11111001_2	$F9_{16}$	9
52_{10}	00110110_2	36_{16}	NBS	105_{10}	10010111_2	97_{16}	p				

EBCDIC CHARACTER CODE GENERIC CLASSES. EBCDIC code has several generic classes of control characters as well as a number of individual characters. Table 2.3 lists EBCDIC's control characters and their descriptions.

TABLE 2.3 EBCDIC Character Codes

Character Codes	Description	Character Codes	Description
ACK	ACKnowledge	IUS/ITB	Interchange Unit Sep/ Intermediate Text Block
BEL	BELl	LF	Line Feed
BYP/INP	BYPass/INhibit Presentation	NAK	Negative AcKnowledge
CAN	CANcel	NBS	Numeric BackSpace
CR	Carriage Return	NL	New Line
CSP	Control Sequence Prefix	NUL	NULl
CU1	Customer Use 1	POC	Program-Operator Comm
CU2	Customer Use 2	PP	Presentation Position
CU3	Customer Use 3	RES/NEP	REStore/eNablE Presentation
DC1	Device Control 1	RFF	Required From Feed
DC2	Device Control 2	RNL	Required New Line
DC3	Device Control 3	RPT	RePeaT
DC4	Device Control 4	SA	Set Attribute
DEL	Delete	SBS	SuBScript
DLE	Data Link Escape	SEL	SELect
DS	Digit Select	SFE	Start Field Extended
EM	End of Medium	SI	Shift In
ENQ	ENQuiry	SM/SW	Set Mode/SWitch

continued on next page

TABLE 2.3   EBCDIC Character Codes (Continued)

Character Codes	Description	Character Codes	Description
ESC	ESCape	SO	Shift Out
ETB	End of Transmitted Block	SOH	Start Of Header
EO	Eight Ones	IT	Indent Tab separator
EOT	End Of Transmission	SOS	Start Of Significance
ETX	End of TeXt	SPS	SuPerScript
FF	Form Feed	STX	Start of TeXt
FS	Field Separator	SUB	SUBstitute
GE	Graphic Escape	SYN	SYNchronous
HT	Horizontal Tab	TRN	TRaNsparent
IFS	Interchange File Separator	UBS	Unit BackSpace
IGS	Interchange Group Separator	VT	Vertical Tab
IR	Index Return	WUS	Word UnderScore
IRS	Interchange Record		

Baudot Code

Baudot code uses 5-bit sequences to define characters. This code, named for its developer Emil Baudot, is one of the earliest codes and is still used by the largest telecommunications networks.

Five bits only provide 32 patterns, which is not sufficient to provide for 26 letters and 10 digits (0—9), spaces and the various control characters like carriage returns, nulls, and more.

To provide more bit patterns, the *shift characters* are used to alert the receiver to interpret subsequent characters using a different character set. There are two character shift methods:

- Figure shifts tell the receiver to use the Figures character set.

■ Letter shifts tell the receiver to use the Letters character set.

Together, these shifts provide 60 characters, as well as control characters in both the Figures and Letters character sets.

BAUDOT, DECIMAL, AND HEXADECIMAL CHART. Table 2.4 illustrates the Baudot Character Code Set and its related decimal and hexadecimal values.

TABLE 2.4

Baudot Character
Code Conversion
Chart

Decimal	Hexadecimal	Binary	Figures Character Code (Shifted)	Letters Character Code (Unshifted)
0	00	0 0000	NU	NU
1	01	0 0001	E	3
2	02	0 0010	LF	LF
3	03	0 0011	A	-
4	04	0 0100	(space)	(space)
5	05	0 0101	S	'
6	06	0 0110	I	8
7	07	0 0111	U	7
8	08	0 0100	CR	CR
9	09	0 1001	D	$
10	0A	0 1010	R	4
11	0B	0 1011	J	BL
12	0C	0 1100	N	,
13	0D	0 1101	F	!
14	0E	0 1110	C	:
15	0F	0 1111	K	(
16	10	1 0000	T	5

continued on next page

TABLE 2.4

Baudot Character
Code Cnversion
Chart
(Continued)

Decimal	Hexadecimal	Binary	Figures Character Code (Shifted)	Letters Character Code (Unshifted)
17	11	1 0001	Z	"
18	12	1 0010	L	)
19	13	1 0011	W	2
20	14	1 0100	H	#
21	15	1 0101	Y	6
22	16	1 0110	P	0
23	17	1 0111	Q	1
24	18	1 1000	O	9
25	19	1 1001	B	?
26	1A	1 1010	G	&
27	1B (figures)	1 1011	SO (shift out)	SO (shift out)
28	1C	1 1100	M	.
29	1D	1 1101	X	/
30	1E	1 1110	V	;
31	1F (letters)	1 1111	SI (shift in)	SI (shift in)

Binary Coded Decimal 8421 Code

The 8421 code is a binary coded decimal (BCD) code that is composed of 4 bits representing the decimal digits 0 to 9. The designation 8421 indicates the binary weights of the 4 bits (see Figure 2.25).

FIGURE 2.25
Binary coded decimal
(BCD) 8421 code.

8	4	2	1
2_3	2_2	2_1	2_0

Two advantages of the 8421 code are that it provides:

- Simple conversion between 8421 code and decimal numbers.

- Data sent in 4 binary bits instead of 8 binary bits, which results saved computer memory.

8421 BCD AND DECIMAL CONVERSION. Table 2.5 illustrates the 8421 BCD values the related decimal values.

TABLE 2.5

8421 BCD
and Decimal
Conversion

8421 (BCD)	Decimal
0000	0
0001	1
0010	2
0011	3
0100	4
0101	5
0110	6
0111	7
1000	8
1001	9

Detecting Errors

Cooperation and synchronization is required between two devices to successfully transmit data. A message is sent one bit at a time from a transmitter to a receiver. The settings on the transmitter and receiver must be the same for the rate, duration, and spacing of bits. The receiver must also be able to recognize the *start* and *stop bit.*

Data transmission is either *asynchronous* or *synchronous.*

- Asynchronous transmission

  — Character transmission is sent one grouping of bits at a time. Most characters are seven bits in length.

  — The receiver resynchronizes at the beginning of each new character, so that timing must only be maintained within each character.

- Synchronous transmission

  — Larger blocks of bits are sent as a unit.

  — The receiver must continuously maintain synchronization with the transmitter.

Facility and terminal equipment is the main contributor to errors, so to ensure reliable data transmission, some type of error detection is required. The basic steps for error detection are:

- Transmitter adds additional error-detecting code to a given frame of bits. This is calculated as a function of the transmitted bits.

- Receiver performs the same calculation.

- If the two calculations compared do not match, the results are considered an error.

The most commonly used error-detecting techniques are:

- Parity Check
- Block Check Character (BOC)
- Vertical Redundancy Check (VRC)
- Longitudinal Redundancy Check (LRC)
- Cyclical Redundancy Check (CRC)
- Echo Checking (Echoplexing)

Parity Check

A parity check is the simplest bit detection scheme. It appends a parity bit to the end of each framed character. To better understand parity checks, let's examine a 7-bit ASCII character. An eighth bit is added to determine proper transmission:

- In synchronous transmission, an odd number of 1s (*odd parity*) is required, as shown in Figure 2.26.

FIGURE 2.26
Odd parity check.

Parity Check bit Odd, Even or Unused		ASCII 7 bit character for the character "N"					
1	1	0	0	1	1	1	0

- In asynchronous transmission, an even number of 1s (*even parity*) is required, as shown in Figure 2.27. An additional 1 is not needed in the parity check field to total an even number of 1s.

FIGURE 2.27
Even parity check.

Parity Check bit Odd, Even or Unused		ASCII 7 bit character for the character "N"					
(blank)	1	0	0	1	1	1	0

Parity check error detection comes into play when the transmitter is transmitting, for example, an ASCII "N" character (1001110_2). Using odd parity, the transmitter appends a 1 and transmits 11001110_2. Upon receipt, the receiver examines the frame; if the total number of bits is erroneously inverted during transmission—i.e., 10001110_2—the receiver detects an error.

Block Check

Parity checking is not always reliable, especially in instances such as noise impulses, which are often long enough to destroy more than one bit. To improve error detection, a second set of parity bits is appended to each character, and a parity bit is generated for each block of characters. The controlling device assigns an additional parity bit for each bit position (column) in the complete frame.

Figure 2.28 illustrates a parity check being determined at the end of each frame of an ASCII 7-bit character, and the parity check being determined at the end of each block of characters. In this example, odd numbers are used.

FIGURE 2.28
Block checking error detection.

Parity bit	ASCII 7 bit							Character
0	1	1	0	0	0	0	1	a
0	1	1	0	0	0	1	0	b
1	1	1	0	0	0	1	1	c
0	1	1	0	0	1	0	0	d
1	1	1	0	0	1	0	1	e
1	1	1	0	0	1	1	0	f
0	1	1	0	0	1	1	1	g
0	1	1	0	1	0	0	0	h
1	1	1	0	1	0	0	1	i
1	1	1	0	1	0	1	0	j
0	1	1	0	1	0	1	1	k
1	1	1	0	1	1	0	0	l
0	1	1	0	1	1	0	1	m
0	1	1	0	1	1	1	0	n
0	1	1	1	0	0	0	0	Block Control Character

Vertical Redundancy Check (VRC) and Longitudinal Redundancy Check (LRC)

Vertical Redundancy Check (VRC) and Longitudinal Redundancy Check (LRC) are used to detect transmission errors on ASCII and EBCDIC character sets:

- ASCII includes:
  - A Vertical Redundancy Check (VRC) is performed on each parity.
  - A Longitudinal Redundancy Check (LRC) is performed on the whole packet.
  - When using VRC and LRC, the block check is a single 8-bit character in the trailer field of the packet.

- EBCDIC includes:
  - No Vertical Redundancy Check (VRC) is made.
  - A Cyclic Redundancy Check (CRC)—16 is calculated for the entire packet. (See the following section for an explanation of Cyclic Redundancy Check [CRC].)
  - Block check is 16 bits long and is transmitted as two 8-bit characters (lowest order transmitted first).

Figure 2.29 illustrates ASCII error detection using VRC and LRC:

- The Vertical Redundancy Check (VRC) is the parity bit at the end of each character that is shown in the row.

- The Longitudinal Redundancy Check (LRC) is the parity check character that is shown in the column.

In both methods, a parity bit is appended to the end of each character as well as to all characters in a frame.

If two bit errors occur in the same column at the same time, errors that bypass the VRC parity check will be detected by the LRC parity check. By using a combination of VRC and LRC, the error rate is reduced greatly over that of just using simple VRC. (The preferred and most commonly used error detection scheme for ASCII 7 is LRC.)

FIGURE 2.29
VRC/LRC error
detection for ASCII.

VRC Parity	ASCII 7 bit							Character	
0	1	1	0	0	0	0	1	a	
0	1	1	0	0	0	1	0	b	
1	1	1	0	0	0	1	1	c	
0	1	1	0	0	1	0	0	d	LRC

Cyclical Redundancy Check

In a switched telephone network, the most common type of error is known as an *error burst*. This burst is caused by a string or burst of consecutive bits in a frame that is corrupted as result of noise impulses caused by the switching elements within exchanges. Parity check does not have the capability to detect error burst, so to fill this void, a technique called Cyclical Redundancy Check (CRC) is needed. The CRC procedure performs the following steps:

- At the end of each message block, the transmitter adds a check character.

- This check character is determined by dividing the message block by a polynomial, discarding the quotient, and using the remainder as the block check character.

- The receiver goes through the same calculation process. When the message block is received, it compares the transmitted block character to its own block check character

- The two possible results are:

  — If the transmitted and received block characters are equal, the message block is assumed to be free of errors.

  — If the block characters are unequal, the receiver makes a request to the transmitter to retransmit the message block.

Echo Checking

Echo checking is mainly used in asynchronous transmission. The control scheme is:

- The receiver obtains a character from the transmitter.

- The receiver immediately *echoes back* the same character to the transmitter for verification.

- The transmitter receives the original character back from the receiver.

- If two characters are equal, the transmitted original character is assumed to be error free. If characters are not equal, the original is presumed to be an error.

- In the case of unequal characters, the transmitter sends to the receiver a control character (such as delete) to ignore the previously transmitted character.

- The receiver performs the necessary deletion and ignores the previously sent character.

- After deletion, the receiver again echoes the character back to the transmitter.

- The transmitter then confirms that the previous character has been ignored.

Because the echo character may have been corrupted during transmission, instead of the originally transmitted character, the echo-checking procedure allows for this possibility to be detected.

Physical Layer (Level 1) Protocols

Telecommunications refers to any communication system that transmits or receives data. Before the transmitter and receiver can exchange data, they must agree on a set of rules between them. Protocols are such sets of rules and exist at many levels, governing physical connections, bit formats into meaningful groups, interpretation of messages speed, service ports, error detection, and error correction.

Figure 3.1 illustrates the placement of the Physical Layer (Level 1) within the telecommunications and OSI Reference Model.

Figure 3.1
Placement of Physical layer (Level 1) in the OSI layered network model.

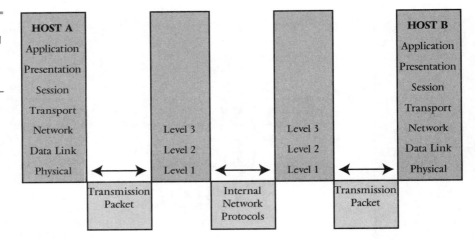

Telecommunications contain levels of protocols that enable proper communication exchange.

- First-level protocols govern:
    — Telecommunication device agreement on which physical medium to use

    — Representation of information on that medium

    — Transmission speed

- Second-level protocols govern:
    — Agreement of how bits will be grouped into meaningful units

    — Detection and correction of transmission errors

- Third-level protocols govern:

    — Agreement of message interpretation between the transmitter and the receiver

Parallel Data Transmission

In parallel data transmission, multiple bits of a data character are transmitted simultaneously over a number of channels. Parallel communication is used in computers; information is exchanged between components along a multiconductor bus, in which multiple bits are placed on the bus conductor simultaneously.

In parallel data transmission over a number of computers, the bits of a data character are transmitted simultaneously over an external connecting line. This method may not be advantageous, however, because there are problems associated with using the parallel connection method between computers:

- Cost

    — Multiple phone connections will be required

    — Multiple repeaters

- Timing or skewing

    — Multiple, long phone lines may cause data transmission to be "off" when compared to signals on the timing lead, which notifies the receiving computer that data is being transmitted

    — Insertion of repeaters into communication path can correct skewing

If speed is not an issue and computers are placed within a few hundred feet of each other, parallel communication may be used. Otherwise, due to cost and skew, parallel communication is never used over public telephone facilities.

Serial Data Transmission

Instead of sending multiple data bits as in parallel communication, serial data transmission sends one bit at a time over a serial line. Serial data transmission is used over telephone facilities. A line interface card is necessary to convert parallel representation of data to serial before such a transmission.

Line Interface Card

A line interface card handles the conversion function required between a data bus and a serial input/output line. It also boosts low voltages. The line interface card contains two components, the Universal Asynchronous Receiver/Transmitter (UART) and a line driver.

- Universal Asynchronous Receiver/Transmitter (UART)
  - Performs tasks while transmitting and receiving asynchronous data:
    - Conversion of synchronous to asynchronous
    - Conversion of parallel transmitted data to serial data
    - Conversion of serial received data to parallel data
    - Bidirectional double buffering
    - Sends and receives start and stop bits
    - Checks parity
    - Communicates with the computer's bus
    - Reports error conditions
- Line Drivers
  - Boost signal levels for distances longer than a few feet
  - Internal line drivers
    - Represents pulses at varying voltages
    - Represents current by turning current on and off in a circuit

- Standards:
    - RS-232-C
    - RS-423-A
    - RS-422-A
    - EIA-232D
    - EIA-232E
- Extended distances
    - Allows DCE to DTE connections over distances greater than 50 feet

For extended distances, a UART may be used by itself within a modem.

Asynchronous and Synchronous Transmissions

There are two types of serial transmission techniques, asynchronous and synchronous:

- Asynchronous transmission (see Figure 3.2) is characterized by:
    - Digital signal transmission
    - Each character consists of only a small number of bits (7 or 8), individually framed by the transmitter
    - No synchronization required between characters
    - No precise clocking required
    - Receiver is relied on to maintain synchronization
    - Baudot code commonly used because it requires fewer bits per character to reduce timing errors
    - Parity bit present (may or may not be used)
    - Various phase relationships and frequencies are contained
    - Transmission of one character at a time

— Encapsulation of individual characters in control start and stop bits that designate the beginning and end of each character

— Character length dependent on the code used

Figure 3.2
Asynchronous
transmission.

Character	Character	Character	Character	Character

- Synchronous transmission (see Figure 3.3) is characterized by:

    — Greater efficiency than asynchronous

    — Digital signal transmission

    — Use of any interface (RS-232, RS-449, EIA-232E, V.345, X.25, etc.)

    — Precise clocking

    — Characters consisting of long sequence of bits sent as a unit, with no breaks between adjacent bits, or sets of bits

    — Synchronization required between characters

    — Same phase relationships and frequencies

    — Whole blocks of data transmitted instead of one character at a time

    — Encapsulation of individual characters in control start and stop bits that designate the beginning and end of each character string

Figure 3.3
Synchronous
transmission.

Character	Character	Character	Character	Character

EFFICIENCY   Asynchronous transmission requires at least two bits of framing for each 8 bits of data. Therefore, it can never achieve better than 80 percent efficiency. Synchronous transmission frames a large number of bits with two bits of framing, so it may achieve a much larger scope of efficiency.

Transmission Flow Categories

The direction of transmission flow is determined by the characteristics of the devices at each end of a channel. When configuring a network device there are three categories of transmission flow directions from which to choose:

- One-way transmission

- Half-duplex transmission (HDX)

- Full-duplex transmission (FDX)

One-Way Transmission

- Transmission is only sent in one direction (Figure 3.4)

Figure 3.4
One-way
transmission.

A   B

- Line has one or more channels

- Direction of transmission flow is determined by the configuration of the devices at either end of the channel

    — Example: We can receive traditional television and radio signals, but are unable to transmit information back

— If a device is configured for one-way transmission and bidirectional transmission is attempted, the data becomes unrecognizable

Half-Duplex Transmission (HDX)

▪ Device at each end of transmission line is configured for bidirection traffic (Figure 3.5)

Figure 3.5
Half-duplex
transmission.

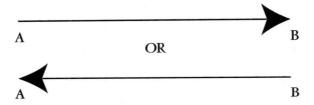

A   OR   B

A   B

▪ Each end device has capability of receiving and transmitting data

▪ Data can only transmit in one direction at a time—not simultaneously

▪ Commonly, half-duplex terminals are connected by two channels to reduce system turnaround time

▪ Also referred to as *either-way transmission system*

— Example: Conversation on telephone—normally, one person speaks at a time; on a CB radio—when pressing the button to speak, you can hear back from base until the button is released

Full-Duplex Transmission (FDX)

▪ Contains two channels for simultaneous data transmission in both directions (Figure 3.6)

Figure 3.6
Full-duplex
transmission.

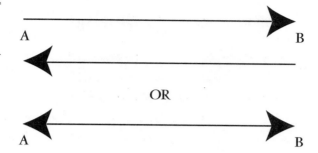

- More efficient than half-duplex transmission because line turn-around time is eliminated

- Consists of two channels, one for each direction

- For full efficiency benefits of full-duplex transmission, both end devices must have the capability of receiving and transmitting data at the same time

- Also referred to as *both-way transmission system*

- Limitations to simultaneous full-duplex can result from competing hardware, terminal equipment, or protocols

- With full-duplex capabilities, reaction time for a system to receive data and retransmit still exists but turnaround time is shorter

Physical Layer (Level 1) Topologies

The Physical Layer specification defines physical connections, signal voltages, and encoding schemes for sending bits across a physical media. It is the lowest layer of the OSI Reference Model.

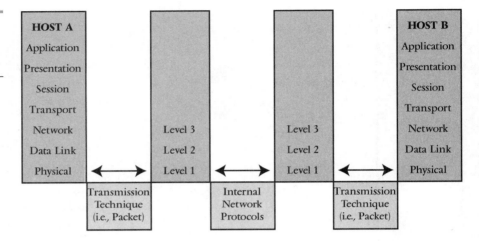

Figure 4.1
Phyical Layer of OSI
Telecommunications
Reference Model.

Physical media encompass the components necessary for transferring signals between systems. Protocols utilized in this layer are responsible for defining the electrical standards and signaling required to generate and detect voltage from data transmission. These protocols describe how to provide electrical, mechanical, operational, and functional connections for telecommunications services. These services are most often obtained from telecommunications service providers, such as regional operating companies, alternate carriers, and post, telephone, and telegraph agencies.

The Physical Layer is responsible for telecommunications equipment. This layer allows data from one system to flow onto another network. To establish this flow of data, each system requires a network connection called a *node*. Nodes must be linked together physically to share network resources. Running wiring between network nodes provides a medium for transmitting data. Most nodes have a physical *Network Interface Card* (NIC) that captures data from the transmitter and translates it to the specifications of the receiver. The physical layout of a network is called its *topology*. The topology of connecting nodes together can be represented in a ring, star, or line pattern. Network wiring and topologies dictate the speed of transmitted data.

Physical Interface Standards

The Physical Layer describes the interface between the *Data Terminal Equipment* (DTE) and the *Data Circuit Terminating Equipment* (DCE). The DTE is a device at the user end of a user network interface that serves as a data source, destination, or both. The DCE provides a physical connection to the network, forwards traffic, and provides the clocking signal used to synchronize data transmission between DCE and DTE devices. Typically, the DCE is the service provider and the DTE is the attached device. Several Physical Layer standards specify the DTE/DCE interface:

- EIA/TIA 232

  — Developed by EIA and TIA

  — Supports unbalanced circuits at signal speeds of up to 64 kbps

  — Resembles V.24 specification

  — Formerly known as RS-232

- EIA/TIA 499

  — Developed by EIA and TIA

  — Faster version of EIA/TIA 232

  — Signal speeds up to 2 Mbps

  — Capable of longer cable runs

- EIA 530

  — Two electrical implementations of EIA/TIA 449

    - RS-422

    - RS-423

- G.703

  — An ITU-T electrical and mechanical specification for connections between telephone company equipment and DTEs

  — Uses BNC connectors

  — Operates at E1 data rates

- V.24

  — An ITU-T standard used between the DTE and DCE

  — Essentially the same as the EIA/TIA-232 standard

- V.35

  — An ITU-T standard describing a synchronous Physical-Layer protocol

  — Used for communication between a network access device and a packet network

  — Most commonly used in the United States and Europe

  — Recommended for speeds up to 48 kbps

- X.21

  — An ITU-T standard for serial communications over synchronous digital lines

  — Used primarily in Europe and Japan

Physical Interface Specifications

These physical interface specifications must be taken into consideration when designing and assembling a physical topology:

- **Wiring or cabling**—Wiring specifications to ensure data delivery at the speed required

- **Connector type**—Specifies the correct connector type, connector size, number of pins, shape of the connector, pin configuration, and any shielding requirements

- **Electrical characteristics**—Specifies required voltage characteristics such as:

  — Level interpretation

  — Acceptance

  — Current level maintenance

  — Signaling type

- — Signal duration
- **Interchange circuit interface**—Specifies circuit relationships, including:
    - — Function
    - — Interactions
    - — Application subset
    - — Circuit subsets for various applications
- **Transmission techniques**
    - — Synchronous or asynchronous transmission
    - — Serial or parallel transmission
    - — Full-duplex, half-duplex, or simplex channel
    - — Dedicated or switched connection
    - — Digital modulation schemes
    - — Analog modulation schemes
    - — Multiplexing scheme

Physical Layer Protocols

The most common Physical Layer protocols include:

- Asynchronous, serial interface
    - — EIA-232-D (formerly RS-232-C)
    - — CCITT Recommendation V.24
- High-speed interfaces
    - — Asynchronous Transfer Mode (ATM) developed by CCITT
    - — High-Performance Parallel Interface (HIPPI) standard, considered part of the ANSI X3T9.3 Committee
    - — Fiber Distributed Data Interface (FDDI)
    - — CCITT Recommendation for X.25: X.21 or X.21 bis

Wiring

Network management and reliability depends on the wiring system utilized. Three types of wiring systems are used in networks:

- Coaxial
- Twisted pair
- Fiber optic

Coaxial Cable

Coaxial cable is a transmission medium that was established by IEEE in 1980. This type of cable is noted for its wide bandwidth and low susceptibility to interference. Construction is made up of an outer woven conductor that surrounds the inner conductor and is separated by a solid insulating material. There are two types of coaxial cable, thick and thin.

THICK COAXIAL CABLE. Thick coaxial cable was the first type of cable used for Ethernet applications. Physical characteristics consist of a relatively large diameter core made out of copper or copper-clad aluminum. The conductor is first surrounded by insulation and then by an aluminum sleeve. The aluminum sleeve is also protected with a polyvinyl jacket. These layers protect the cable from environmental assaults and allow dependable transmission, but on the negative side, thick coaxial cable is difficult to bend because of the large diameter of the copper conductor. Because of its thickness and the difficulty of manipulating it, thick coaxial is not as popular as other cabling.

Thick coaxial cable works on bus networks using transmission speeds of 10 Mbps. According to IEEE standards, the maximum cable length is 500 meters. The shorthand for these specifications is 10BASE5. The 10 indicates the cable transmission rate is 10 Mbps. BASE means that baseband transmission is used. The 5 indicates 5 × 100 meters for the longest cable run.

To accurately determine the placement of network devices, the cable jacket is conveniently marked every 2.5 meters. If devices are not attached at these appropriate distances, network errors may result.

The IEEE specification states that by using the addition of repeaters to the cable run, the maximum cable length can be increased to 2,500 meters, which will support up to 100 nodes.

Thick coaxial cable can be used for two different types of transmission modes:

- **Baseband**—Used on Ethernet networks and supports the CSMA/CD access method. Baseband Ethernet packets travel at a percentage of the speed of light, known as the *propagation velocity* (V_p). The *Maximum Medium Delay* (MMD) is determined by the media type, V_p, the number of devices, and the segment length. For thick coaxial cable, the Maximum Medium Delay per segment is 2,165 nanoseconds (ns).

- **Broadband**—Consists of a set of distinct channels. Each channel operates at a unique frequency. The broadband mode was designed to support diverse signal transmissions (cable TV, data, etc.).

THIN COAXIAL CABLE. Thin coaxial cable is a baseband cable that supports a data rate of 10 Mbps on Ethernet bus topologies. The IEEE standards for this type of cable were established in 1985. Thin coaxial cable is also known as RG-58. Maximum cable segment length is 185 meters, with up to 30 nodes per segment. Shorthand notation for this type of cable is 10BASE2. Maximum Medium Delay per segment is 950 ns. Segments are marked every 0.5 meters for attachment of devices.

Physical construction of thin coaxial cable is similar to that of thick coaxial cable, except that the diameter of the cable is significantly smaller than that of thick coaxial. A copper conductor is at the center of the cable, surrounded by insulation that is surrounded by a woven mesh outer conductor. The woven mesh outer conductor is then surrounded by a polyvinyl jacket.

Thin coaxial cables have a thin diameter and flexibility that make it suitable to run through walls and ceilings.

Twisted-Pair Cable

Twisted-pair is a type of cable in which pairs of conductors are twisted together to produce certain electrical properties. These pairs of conductors are also referred to as *two-* or *four-wire circuit* (wires must be even numbers). The circuits formed by the two conductors are insulated from each other. There are three types of twisted-pair cable:

- Unshielded twisted-pair cable

- Ethernet twisted-pair cable

- Token-ring applications of twisted-pair cable

UNSHIELDED TWISTED-PAIR CABLE.   Unshielded twisted-pair (UTP) cable is telephone wire. IEEE formalized the cable for networking in 1990. It is thin and pliable so that it is easy to install, costs much less than coaxial cable, and permits the use of existing telephone cable that was installed within the last 5 to 10 years. The IEEE shorthand for UTP cable is 10BASET.

UTP consists of four individual strands of wire that have several twists per foot of cabling. These twists ensure that the electrical signal is not attenuated. The ends of each cable run are attached to RJ-45 connectors.

ETHERNET TWISTED-PAIR CABLE.   Ethernet twisted-pair cable is used for Ethernet bus applications where there is no external terminator. The cable is terminated within the hub and the Network Interface Card (NIC). Transmission speed on the cable equals 10 to 100 Mbps. Maximum segment length from the hub to the workstation is 100 meters. The Maximum Medium Delay per segment is 1,000 ns.

Ethernet twisted-pair cable is used in a physical star configuration, with one node per segment. These nodes are connected via a *concentrator.* Troubleshooting is easy because it is simple to trace a bad node or wire run, making faulty installation problems easier to trace than when using thick or thin coaxial cable.

The availability of UTP for Ethernet networks has opened the way for vendors to develop 100 Mbps Ethernet communication rates. The IEEE endorses using 100BASE-VG to accommodate these higher speeds.

TWISTED-PAIR CABLE FOR TOKEN RING APPLICATIONS. Token ring networks can use shielded or unshielded twisted-pair cable. Shielded twisted-pair has shielding around the strands of cable to help reduce outside interface. Three types of wire are normally used for token ring:

- Type 1 wire is shielded twisted-pair:

  — Consists of two solid wires surrounded by shielding

  — Braided shielding is used for indoor applications

  — Corrugated metallic shielding is used for outdoor applications

  — When a single MAU (central communications hub) is present, the maximum cable segment is 300 meters

  — When multiple MAUs are installed, the maximum cable segment is 100 meters around the shield

  — A DB-9 connector is used at the workstation or node end

  — At the MAU end, a hermaphroditic connector is used

- Type 2 wire is shielded twisted-pair cable:

  — Consists of two solid twisted-pair cables in the middle, shielding over the middle wires, and four solid twisted-pair cables around the shield

  — Braided shielding is used for indoor applications

  — Corrugated metallic shielding is used for outdoor applications

  — A DB-9 connector is used at the workstation or node end

  — At the MAU end, a hermaphroditic connector is used

- Type 3 wire is unshielded cable:

  — Used in telephone cable plant in a building

  — A media filter is placed at each network node to filter out noise or undesired signals on the wire

  — Maximum cable segment for type 2 and 3 wire is 100 meters if used with one MAU

  — Uses RJ-11 and RJ-45 connectors

Fiber Optic Cable

Fiber Distributed Data Interface (FDDI) applications use fiber optic cable in a dual ring configuration. This cable is composed of a central glass cylinder that is encased in a glass tube, called *cladding,* which is surrounded by a polyvinyl cover. The cable core carries optical energy as transmitted by *laser* or *light emitting diode* (LED) devices. The glass cladding is designed to reflect light back into the core.

Fiber optic cable is well suited for FDDI, Fast Ethernet, SONET, and ATM networks for several reasons:

- Capability of propagating transmitted light wave at high speeds

- Supports high bandwidth with low attenuation over long distances

- No outside interference problems

- High security against wire tap

A negative note about fiber optic cable is that because of its glass construction, it is very fragile.

The success of data transmission by light waves is determined by the wavelength of the light. Data transfer does not travel efficiently over visible light; however, infrared light provides the necessary efficiency for data transmission.

Power loss on fiber optic cable is measured in decibels (dB). It can be the result of the length of the cable or passage through connectors and splices. The maximum attenuation for FDDI applications is 1.5 dB/Km. The minimum power level required for data transfer on fiber optic cable is called the *power budget.* For FDDI communications, the power budget must by 11 dB. The maximum segment length is 1,000 meters, and the Maximum Medium Delay per segment is 5,000 ns.

SINGLE-MODE AND MULTIMODE FIBER OPTIC CABLE.
Fiber optic cable comes in two modes:

- Single-mode cable is used mainly for long distance communications

  — Central core diameter is much smaller than that of multimode cable because it only carries one transmission

- — Laser light is the communication source
- — Laser light source, coupled with a relatively large bandwidth, enables long distance transmissions at high speeds
- ▪ Multimode cable is used mainly for local communications
  - — Central core diameter is larger than that of single-mode to support simultaneous transmission of multiple light waves
  - — Available bandwidth for local communications is smaller and the light source is weaker so that the distance is shorter than in the single-mode cable
  - — A light-emitting diode (LED) is the transmission source

Physical Layer (Level 1) Signaling

Signaling is described as any invisible air vibration that stimulates the auditory nerves and produces the sensation of hearing. These vibrations take the form of a *sine wave* which radiates in all directions from the source.

Bandwidth measures the range that a normal human ear is capable of hearing. The lowest range is approximately 20 Hz and the highest range is approximately 20 KHz. In telecommunications, it is not necessary to use the entire bandwidth. Clear signals are possible within the 200 to 3,400 Hz range.

The process used by a device (transmitter) to send information or receive information (receiver) over a telephone line is referred to as *communication*. There are two forms, voice and data.

To transfer voice or data, the telephone facility between the transmitter and receiver must convert the information to electrical signals, then convert these signals back to their original form at the receiving end. The electrical signal is represented by an *alternating current (AC) sine wave*, which is broken down into units called *cycles*. One cycle is determined by measuring between two identical points in the AC sine wave (Figure 5.1).

Figure 5.1
Sine wave.

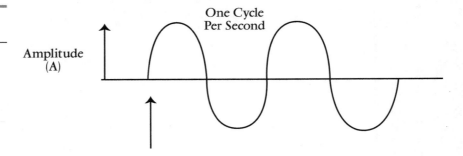

Data, defined by a telephone company, is the transfer of information from one location to another over telephone lines and equipment by means other than the human voice. Data is transferred through two types of electrical signals, analog and digital.

Analog Signal Waves

An analog signal takes the form of a continuously varying physical quantity of electrical signals. *Analog data* is defined as information that can assume an infinite number of values during a specific time frame. The shape of an analog signal is the same as the curved or circular sine wave it represents. Sine waves are described by three properties or parameters:

- Amplitude
  - Height of a wave above (or below) the axis
  - Level, strength, volume, or loudness of the signal
  - Measurement values expressed in decibels (dBs)
  - Reference level for amplitude is a fixed value of 0 dBm, which is equal to one milliwatt of power

- Frequency
  - Number of times (cycles) the object moves around the circle per unit time
  - Period of time is usually expressed per second
  - Measurement values are expressed in Hertz (Hz)
  - Described in ranges called bandwidth

- Phase
  - Number of degree differences between identical sine waves at identical points in the sine wave
  - Relative measurement of horizontal axis crossings
  - Measurement values are expressed in degrees

To graphically represent a sine wave, the height of a signal moving around a circle at a constant rate is plotted on a *y*-axis versus the distance traveled on an *x*-axis. The three parameters of the sine wave can be varied systematically to represent data.

Figure 5.2 illustrates a varying phase where two sine waves have the same amplitude and frequency, but start at different times.

Figure 5.2
Varying phase when two sine waves hae the same amplitude and frequency.

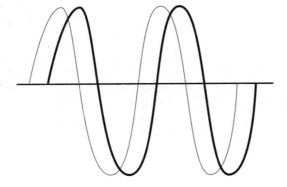

When the amplitude varies as shown in Figure 5.3, the height (power) changes and crossing points of the sine wave remain constant, if phase and frequency are the same.

Figure 5.3
Varying amplitude in sine waves having the same phase and frequency.

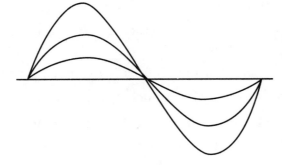

When the frequency is varied as shown in Figure 5.4, the point at which the sine wave crosses the zero reference changes, but the amplitude remains consistent.

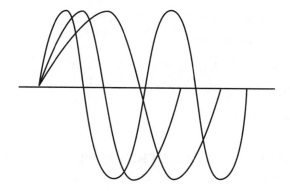

Figure 5.4
Varying frequency in
a sine wave when
amplitude and phase
remain constant.

Analog Transmission

If the analog signal between telecommunications equipment is not a direct connection and must be transmitted through a telephone line, it is necessary for a device to convert the analog signal to one consistent with the signal required by the communications facility. When analog and digital circuits interact, the two signals conflict. To correct this conflict, *Network Channel Terminal Equipment* (NCTE) and *modems* are designed into the circuit.

The modem is an analog-to-digital and digital-to-analog signal converter that works by modulating a signal onto a carrier wave at the originating end and demodulating it at the receiving end. The word modem is derived for MOdulation and DEModulation.

- **Modulation** converts a communication signal from analog to digital or digital to analog for transmission over a medium between two locations.

- **Demodulation** converts the communication signal back to its original format.

Analog Modulation

There are six types of analog modulation:

- Amplitude Modulation (AM)

- Frequency Modulation (FM)

- Phase Modulation (PM)

- Frequency Shift Key (FSK) Modulation

- Bit Rate

- Baud Rate

AMPLITUDE MODULATION (AM). Amplitude modulation (AM) (Figure 5.5):

- Represents the difference between the most positive voltage and the most negative voltage

- Represents the lowest and highest points of the sine wave

- Shows changes in height of the sine wave to show changes in the information signal

- Has noise interference that changes the amplitude level of the signal, making it impractical for use alone in data communications

- Is most commonly used in radio broadcasting

Figure 5.5
Amplitude Modulated (AM) signal.

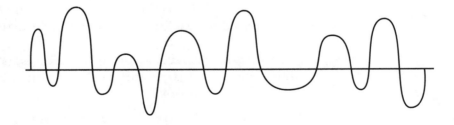

FREQUENCY MODULATION (FM). Frequency modulation (FM) (Figure 5.6):

- Represents the number of cycles in a given period of time

- Shows changes in the number of cycles during a period of time to show the changes in the information signal

- Is a "quieter" signal than AM

- Minimizes the interference from noise, which affects amplitude rather than frequency

- Is most commonly used in radio broadcasting

Figure 5.6
Frequency
Modulation (FM)
signal.

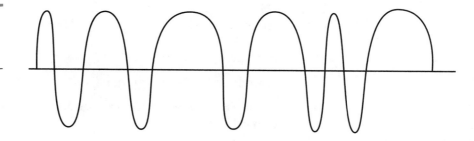

PHASE MODULATION (PM). Phase modulation (PM) (Figure 5.7):

- Represents the number of degrees difference that two sine waves are offset from one another

- Shifts a sine wave 180 degrees whenever the bit stream changes from 1 to 0 or 0 to 1

- Is not used for broadcasting

- Is less capable of transmitting small changes in the modulating signal than AM or FM

- Has more transitions between specific states

- Works well for data transmission

Figure 5.7
Phase Modulation
(PM) signal.

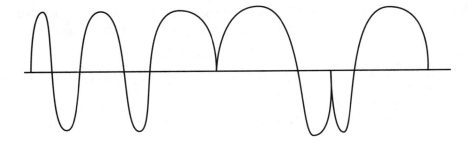

FREQUENCY SHIFT KEY MODULATION (FSK). Frequency shift key modulation (FSK):

- Is an offspring of FM

- Uses speeds of 300 b/s or higher

- Shifts the frequency of the carrier according to the state of the pulses modulating it

- Responds to customer equipment sending a "0" bit by signaling the modem to turn on an oscillator that sends an analog wave of a specific frequency

- Responds to customer equipment sending a "1" bit, by signaling the modem to turn on another oscillator that generates a different frequency

BIT RATES. To review, a bit is defined as a component of information that can assume only certain distinct values or patterns during any specific time. The bit rate refers to how many bits can be transmitted in one second.

- Bit rate can be 2,400 bits per second (bps) or higher

- Common references to the bit rate are:

  — 2.4 Kbps

  — 64 Kbps

  — 1.544 Mbps

BAUD RATES. Baud rate is:

- The number of times the line condition changes per second

- Not a specific reference to the operation speed of the customer's equipment

- Equal to the bit rate if one bit is sent with each analog signal change (i.e., if the bit rate is 6.4 Kbps and the baud rate is 6.4 Kbps changes per second, they are equal.)

- Not equal to the bit rate if the change in the line signal represents more than a single bit (i.e., if the bit rate is 1.544 Kbps and the baud rate is 6.4 Kbps changes per second, then the modem encodes 4 bits at a time [$1.544 \times 4 = 6.4$].)

Digital Transmission

Digital transmission is the most common means of sending or receiving data. Digital data is defined as information that can assume only certain distinct values or patterns during any specific time. Two examples of digital data are the dial pulse and Morse code.

The analog waveform has three basic characteristics, whereas the digital pulse has six basic characteristics. Figures 5.8 through 5.14 illustrate various characteristics of digital signals and their definitions.

Figure 5.8
Digital pulse
characteristics.

Frequency   Amplitude   Duration   Shape   Position   Polarity

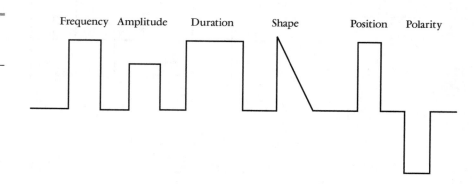

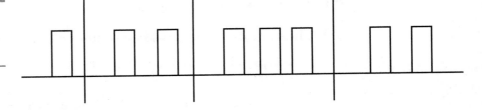

Figure 5.9
Frequency (number of pulses per second).

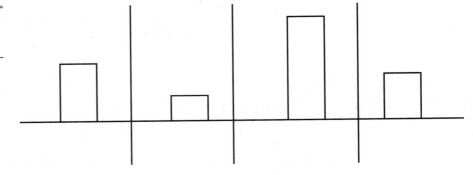

Figure 5.10
Amplitude (height of the signal).

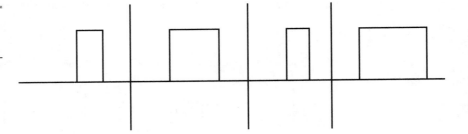

Figure 5.11
Duration (length of time for the pulse).

Figure 5.12
Position (location of the pulse, which can start at the beginning, middle, or end of the time slot).

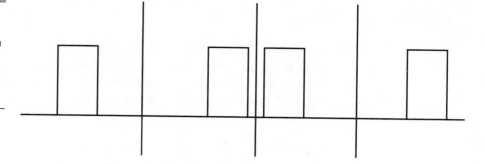

Figure 5.13
Shape (measured by change in rise time and decay time).

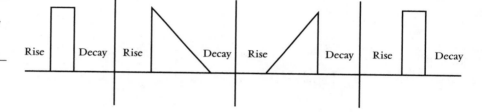

Figure 5.14
Polarity (determined by whether the pulse is positive or negative in relation to the reference point).

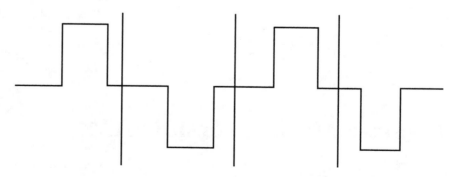

Digital Signal

A digital signal represents information with a code that changes in one or more characteristics. Unlike the analog signal, which represents information in variable but continuous waveforms, the digital signal is discontinuous in time. There are two types of digital signals (see Figure 5.15):

- **Unipolar**—From "uni" meaning "one." A unipolar signal consists of only one voltage polarity that is either positive or negative.

- **Bipolar**—From "bi" meaning "two." A bipolar signal consists of both positive and negative polarities.

Figure 5.15
Unipolar versus
bipolar digital signals.

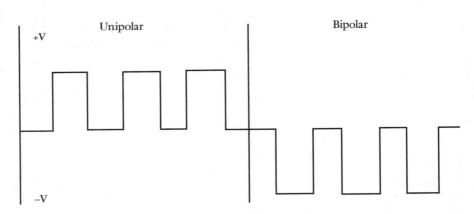

Alternate Mark Inversion (AMI)

Alternate mark inversion (AMI) is a bipolar signal coding zeros for the absence of a pulse and ones for the presence of a pulse. The ones are coded alternately as positive-going and negative-going pulses (Figure 5.16). Positive-going pulses have a positive voltage reading, whereas negative-going pulses have a negative voltage reading.

Figure 5.16
Alternate Mark
Inversion (AMI).

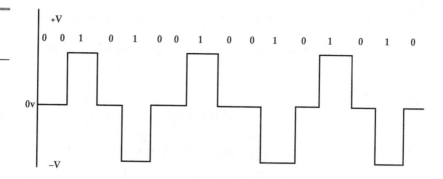

Advantages of AMI over unipolar and bipolar pulse coding schemes are:

- Error detection
- No direct current (dc) component
- Reduced bandwidth requirement

Error Detection

AMI has an error detection method call *bipolar violation detection*. This method detects *bipolar violations* (BPVs). Because AMI has alternating positive and negative pulses, an error is detected when two consecutive pulses of the same polarity occur—either two negative pulses or two positive pulses together (Figure 5.17).

Figure 5.17
Bipolar violation
detection.

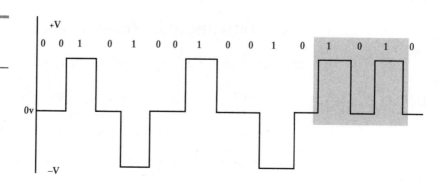

No Direct Current (dc) Component

AMI does not require a dc component, so it's a much cleaner and more efficient method of transmitting data the unipolar method (Figure 5.18).

■ A unipolar signal may start at 0 volts dc but each pulse rises to some specific dc value. Capacitance causes the voltage to continue to rise rather than return to zero. This called the *dc component*.

■ The AMI signal also starts at 0 volts dc, but because each "one" bit is of opposite polarity, the positive (plus) and, negative (minus) dc components cancel each other out.

Figure 5.18
AMI and unipolar
signal comparison.

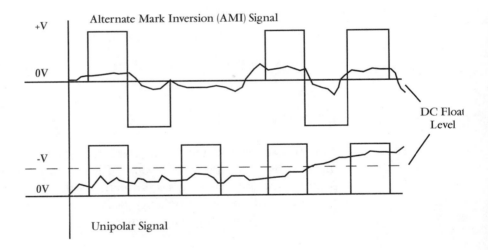

Reduced Bandwidth Requirement

Alternate mark inversion (AMI) is much more efficient than unipolar transmission because it requires less bandwidth to transport data. The frequency of a unipolar signal is twice the *frequency* of an AMI signal carrying the same amount of bits. Because the AMI signal uses the positive pulse of the cycle to carry one bit and the negative pulse of the same signal to carry another bit, the AMI signal reduces the bandwidth to half that of the unipolar signal (Figure 5.19).

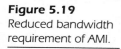

Figure 5.19
Reduced bandwidth
requirement of AMI.

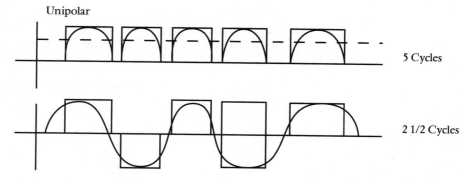

Transmission Impairments

The ideal transmission path is supposed to deliver an accurate reproduction of the original signal to the receiving end. Analog transmission signals are rounded (sine wave) and digital signals are square wave pulses that are transmitted over the telephone lines. In digital transmissions, the signal is transmitted by the presence or absence of the pulse and not by the shape of the pulse.

Impairments to digital transmission include

- Loss

- Noise

- Distortion

Loss

Loss, also know as attenuation, is caused when the output level is less than the input level (i.e., a tone is sent at 0 dBm level and received as negative dBm). The two most common methods to compensate for loss are *amplification* and *regenerative repeating*.

- **Amplifier**—An amplifier is a device used to compensate for transmission loss. When an amplifier is inserted in the circuit, it increases the power or adds gain to the circuit. Gain is the result when transmission output is greater than transmission input. This boost to the amplitude of the entire input signal includes any existing noise. In schematics, an amplifier is represented by a triangle (Figure 5.20).

The point of the triangle always points in the direction of the signal.

Figure 5.20
Amplifier.

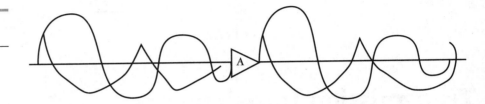

- **Regenerative Repeater**—A regenerative repeater generates a new pulse if the input signal meets or exceeds a designed threshold level. If the input signal does not meet or exceed the threshold level, no new pulses are generated (Figure 5.21).

Figure 5.21
Regenerative
repeater.

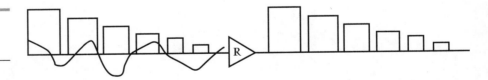

Noise

Noise is unwanted electrical signals that interfere with the information signal.

- Noise develops during transmission

- Induction from power lines is the most common source of noise, but it may also be the result of radio interference, weather, etc.

- Noise sources are cumulative

- Noise is the biggest problem in telephone circuits

- Noise is measured in decibels (dB above reference noise [rn] or dBrn)

- —90 dBm is the reference noise (zero noise) level

- Message circuit noise

  — Noise that lasts 200 milliseconds or longer

  — Present at all frequencies

  — Amplitude that can constantly change

- Impulse noise

  — Noise that lasts less than 200 milliseconds in duration

  — Only important in data circuits and not voice circuits

Distortion

Distortion is any change in the waveform of a signal that occurs while it is being transmitted over a telephone circuit. Distortion is not acceptable to users at any level. There are two types of distortion:

- Attenuation distortion

  — Various frequencies within a specified band are not attenuated equally

  — Higher frequencies are most distorted

  — Common cause is capacitance in the cable

  — The longer the cable, the more distortion increases

- Delay distortion

  — Result of frequency components of a signal traveling slowly over a transmission path

  — Detrimental to data transmissions

  — Little effect on voice transmission

Telecommunications Signaling Applications

There are two types of signaling applications used in the telecommunications facility environment:

- Direct current (dc)
- Alternating current (ac)

Direct Current (dc)

Direct current (dc) signaling application formats are :

- Loop Start
  - Primarily used for residential and business telecommunications.
- Ground Start
  - Used by PBX customers.
- E & M
  - A limited range as low as 50 ohms makes E & M more useful in electromechanical rather than electronic switching systems.
- Duplex (DX)
  - Extends E & M by up to 5,000 ohms using *simplex resistance*.

LOOP START. Loop start (Figure 5.22) is a signaling application primarily used for residential and business customer service.

This application is a telecommunications facility that provides —48V on the *ring* and a ground on the *tip* of the cable that produces a two-state signaling process. A simple closure of the loop between the tip and ring is needed to start current flowing.

Figure 5.22
Loop start.

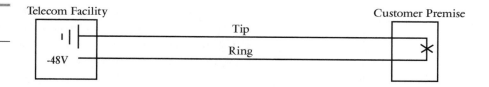

GROUND START. Ground start (Figure 5.23) is a signaling application primarily used in Private Branch Exchange (PBX) systems. This application is a four-state signaling process that has switching equipment at both ends of the loop to control the use of a trunk. When idle, the tip is open in the telecommunications facility; upon seizure, a ground is applied.

Figure 5.23
Ground start.

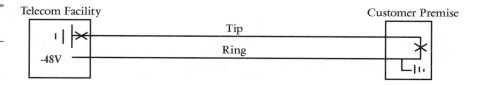

E & M SIGNALING. E & M signaling is a signaling application which contain a limited range as low as 50 ohms. E & M Signaling circuits contain two leads—one lead for the "M" lead function and one for the "E" lead function.

The "M" lead seizures have battery connection.

The "E" lead is open in idle states and has a ground state when seized.

Both leads are attached with a common ground path for current flowing between switching and signaling equipment. Each lead is used in opposite transmission flow directions to produce one of four:

- Ringing—no ringing
- On-hook—off-hook
- Idle—seized
- Tip ground—tip open

The equipment to which the E & M leads are attached determines the state. *Source* equipment is at the changing end; in *nonsource* equipment, the opposite end of each lead has a sensor that never changes.

There are two types of E & M Interfaces

- Type I
- Type II

E & M TYPE I INTERFACE.   E & M Type I Interface is a two-wire interface. This traditional type of interface has one lead for the "M" lead function and one lead for the "E" lead function. The "M" lead uses nominal —48V for seizure and has a ground in the idle state. The "E" lead uses ground from the signaling equipment for seizure and has an open in the idle state (Figure 5.24).

Figure 5.24
E & M Type I
interface.

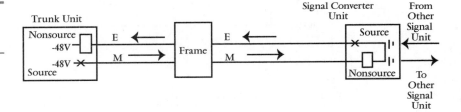

Type I was the original interface used for *step-by-step* and *crossbar-type* switching machines. Although E & M signaling circuits perform well in electromechanical switching systems, they do not always provide satisfactory performance in the *Electronic Switching Systems* (ESS). Type II would be our next step.

E & M TYPE II INTERFACE.   E & M Type II Interface is a four-wire, fully looped arrangement that uses open and close signals in each direction between the trunk circuit and the signaling circuit. Unlike the Type I interface, this closure is across two leads (M and SB) instead of one, thus permitting the ground and battery to come from the same source.

Figure 5.25
E & M Type II
interface.

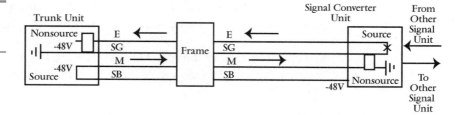

In Figure 5.25:

- The trunk circuit signals *idle* by opening the relay contact between the "M" and "SB" leads and signals a *seizure* by closing the contact between the "M" and "SB" leads.

- Signal converter circuit signaling to the trunk circuit is achieved by the same opens and closures between the two "E" and "SG" leads.

- The relay contact in the signaling *idle* circuit is open between the "E" and "SG" leads.

- The relay contact in the signaling *seizure* circuit is closed between the "E" and "SG" leads.

- The trunk circuit supplies the ground for the "SG" lead, which provides the "E" lead with an open for an idle and a ground for a seizure.

DUPLEX (DX).   E & M signaling has range so that it becomes necessary to use a duplex signal converter to allow E & M to signal over cable loops in excess of 50 ohms. A duplex unit allows E & M signaling to be extended over loops having up to 5,000 ohms of *simplex resistance.*

DX signaling circuits use two- or four-wire line facilities composed of cable pairs equipped with repeating coils at both ends. Because duplex signaling is limited to 5,000 ohms simplex resistance, four-wire loops are more commonly used because the resistance is half that of a two-wire loop, thus allowing twice the distance.

A DX data signaling circuit uses the same conductors as the voice path without creating any interference and thus eliminating the need for additional cable facilities.

For the duplex converter to function properly, the equipment must be in balance. Balance is defined as the state wherein:

■ The circuit must have a DX unit located at the telecommunications facility and an exact copy of the DX unit at the customer premise.

■ The internal balance of the DX network and the customer network? are adjusted so that their resistance is equal to each other.

The DX signaling converter has two leads, "A" and "B". These leads connect internally to the duplex—E & M converter and are simplexed to the repeat coils. This permits the use of the same pair for signaling and transmission.

■ The "A" lead is the *supervisory* or *signaling* lead, which provides seizure (−48 VDC) and idle (−2 VDC) signaling states in both directions

■ The "B" lead is the *bias* or *balance* lead, which provides about −20 VDC. This lead is used to compensate for the ground potential differences between the two ends of the circuit. It functions to keep or force the *polar relay* in the converter into a "biased" or released state.

POLAR RELAY. The polar relay device allows current flow in one direction. When it receives current flow in the opposite direction, the relay opens or releases. Polar relays consist of one moving contact, which is used as a break contact, and two fixed contacts used as a make contact set.

Polar relays are effective in dial pulses transmission and are used in duplex—E & M converters.

Operation of the polar relay is illustrated in Figure 5.26. The winding polarity is shown as ±. The polar relay goes through a process called biasing the relay: If the polarities match ("A" contacts are closed), the polar relay operates. If they do not match ("B" contacts are closed), the relay is forced to a released or biased state.

Figure 5.26
Polar relay.

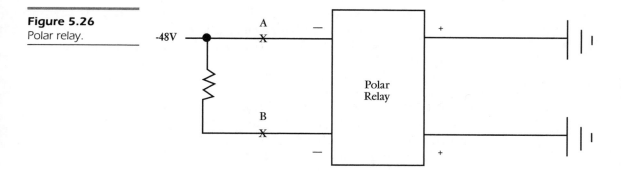

Figure 5.26
Polar relay.

DUPLEX (DX) INTERFACES

- **DX I, Type I Interface**—This is a device used to extend the E & M signaling range. It requires an "M" lead input on the E & M leads. A DX I, Type I Interface is used to call the trunk circuit to provide battery for seizure and a ground for idle on the "M" lead to the duplex—E & M converter. In response, the signaling converter supplies a ground on the "E" lead for seizure.

- **DX II, Type I Interface**—The DX II, Type I Interface is a device used to extend the E & M signaling range. It requires an "E" lead input on the E & M leads. A DX II, Type I interface calls for the application of a ground on the "E" lead to produce a seizure. The trunk circuit supplies the ground. In response, the signaling converter supplies a ground for the "M" lead operation.

- **DX I, Type II Interface**—The DX I, Type II Interface is a device also used to extend the E & M signaling range. This device requires an "M" lead input on the E & M leads. This interface requires battery on the "M" lead for seizure. The signaling converter supplies battery on the "SB" lead through a closure in the trunk circuit to the "M" lead. The trunk circuit supplies a ground on the "SG" lead through a closure in the signaling circuit to the "E" lead.

- **DX II, Type II Interface**—The final arrangement of duplex devices is the DX II, Type II Interface, which is also used to extend the E & M signaling range. This device requires an "E" lead input on the E & M leads. A DX II, Type II interface requires a ground on the "E" lead for seizure. The signaling circuit will supply ground on the "SG" lead through a loop closure in the trunk circuit to the "E" lead. In response, the trunk circuit supplies battery on the "SB" lead through a closure in the signaling circuit to the "M" lead.

A & B LEAD REVERSAL.   The simplexing arrangement provided in duplex—E & M equipment, whether in a telecommunications facility or in network channel terminating equipment, provides a simplex reversing switch (Figure 5.27). This switch permits the reversing of the simplex leads at one end of the circuit so that both "A" leads are connected to the *supervision* path and both "B" leads are connected to the *bias* path.

Figure 5.27
A & B lead reversal.

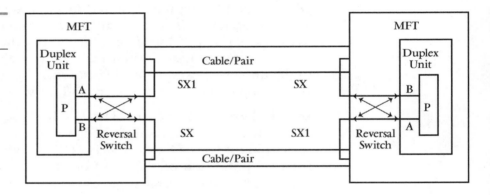

As a note, the proper operation insurance must be provided:

- The internal "A" and "B" leads of the duplex signaling unit must be properly optioned.

- DX converters permit two states of operation over the "A" lead in both directions, seizure and idle.

- The "A" and "B" leads inside the two DX converters must be *normal* on one end and *reversed* on the other end.

- The DX converter at each end of the circuit must be balanced.

- Resistance and the signaling path must be equal to each other

- The "A" lead of one DX unit must be connected to the "A" lead of the other unit to ensure that the internal "A" and "B" leads are properly optioned.

Alternating Current (ac)

Alternating current (ac) signaling systems provide a means to convey supervision and address information over a transmission facility that exceeds the range of direct current system. Two ac signaling formats are:

- Multifrequency (MF)

- Touch Tone (TT)

MULTIFREQUENCY (MF). Multifrequency (MF) ac transmission is a system of sending address signals using the transmission path of the outgoing and incoming trunk circuits between telecommunication facilities. These signals are either

- Modulated

- Demodulated

Modulated frequencies are sent together over a telecommunications facility to be demodulated, detected by the multifrequency (MF) receiver at the opposite end, and recorded as a digit. Table 5.1 illustrates multifrequency numeral and frequencies relationships.

TABLE 5.1

Multifrequency Numeral to Frequency Relationships

Numeral	Frequencies
1	700 + 900
2	700 + 1100
3	900 + 1100
4	700 + 1300
5	900 + 1100

continued on next page

TABLE 5.1

Multifrequency
Numeral to
Frequency
Relationships
(Continued)

Numeral	Frequencies
6	1100 + 1300
7	700 + 1500
8	900 + 1500
9	1100 + 1500
0	1300 + 1500
ST	1100 + 1700
KP	1500 + 1700

TOUCH TONE (TT). Touch Tone (TT) ac transmission is very similar to multifrequency (MF) except that TT uses a combination of two frequencies to send address signals from the customer's telephone to the telecommunication's switch.

Figure 5.28 shows a common customer telephone Touch Tone keypad indicating the amount of frequency required for number translation.

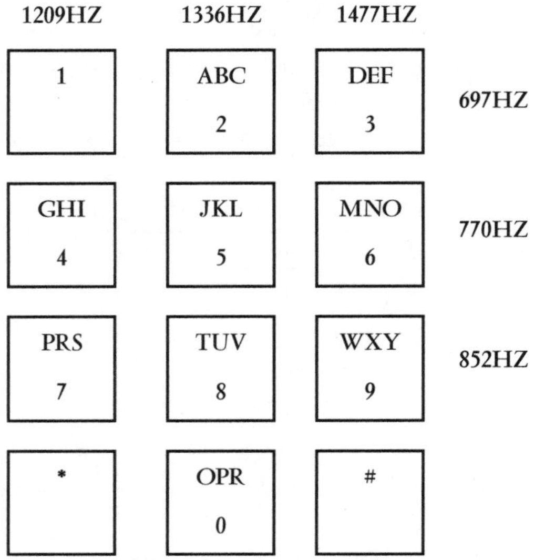

Figure 5.28
Touch tone keypad
and corresponding
frequencies.

Data Link Layer
(Level 2)

The Data Link Layer is the second lowest level of the OSI Reference Model. It defines standards for dividing data into packets and sending the packets across the network.

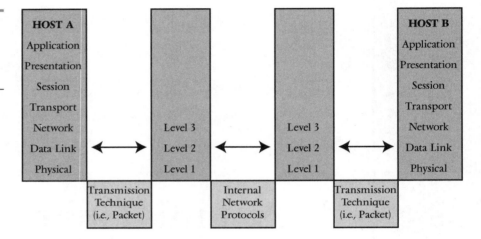

Figure 6.1
Data Link Layer of the OSI Telecommunications Reference model.

Data packets are constructed at this level, with each packet formatted for the synchronization of data transmission. The Data Link layer also creates the formatting that allows packets to contain physical addressing.

The Data Link Layer is responsible for providing error-free reliable data transmission from one network node to another or between two adjacent devices. Telecommunication is first established by transmitting of a set of signals. As soon as the connection has been made, signals are formatted into packets. Data is encoded in an electrical signal by the transmitting device, decoded by the receiving device, and checked for errors. Once communication is verified between two devices, their Data Link Layers are connected physically (through the physical layer) and logically (through peer protocols).

In IEEE 802, the Data Link Layer incorporates the *Logical Link Control* (LLC) protocol. LLC is an example of a peer protocol. It enables two communicating Data Link Layers on separate nodes to have common guidelines for flow control, error handling, and data retransmission.

Protocols for Encoding and Decoding Data

A *protocol* is a set of specific rules that terminals at each end of a transmission line are required to follow when converting a received or transmitted message into a serial bit stream. A standard set of basic functions that any data link protocol must perform to be considered a true data link protocol includes:

- Connections for use by the Network Layer
- Establishment of connection with one or more physical nodes
- Delimiting of data so that it can be transmitted as frames
- Synchronization of split data upon receipt
- Performance of a minimum set of control functions
- Transmission of frames marking the beginning and ending of each transmission frame

With new advanced technology, some data link protocols are capable of performing these additional functions:

- Addressing (network address manipulation)
- Pacing (transmission rate control when data is transmitted faster than the receiver can handle it)
- Retransmission of frames (resending correct versions of frames that have errors)
- Status inquiry (control functions allowing one device to inquire about the status of another)

Because the Data Link Layer is concerned with the error-free transmission of data, this layer checks incoming signals for duplicate, incorrect, or partially received frames of data. If an error is detected, a retransmission of the data is requested. As the Data Link Layer transfer frames up to the next layer, it ensures that frames are sent in the same order as received.

The following list of high-level services that the Data Link Layer provides will be described in greater detail as each functional is reviewed:

- **Connection**—Connections are established and released dynamically between two or more physical nodes as defined by the network layer.

- **Connection endpoint identifiers**—Defined connection endpoint identifiers are used by the Network Layer.

- **Timeout**—Procedures must exist to ensure that a response to frames is received within a specified period of time.

- **Addressing**—The transmitter and/or receiver of a frame must be specified.

- **Framing**—A frame is a specific sequence of bits of data. Framing marks the beginning and end of a transmission frame.

- **Sequencing**—Frame delivery must be in the same order as transmitted.

- **Parameter**—Parameters define the quality of service and include errors, error rate, availability, delay, and throughput.

- **Control**—Specifies a transmitter's ability to identify the receiving machine.

- **Line control**—If the channel is half-duplex, procedures must exist to determine which station on the line may transmit.

- **Flow control**—Ensures that a transmitter does not transmit more frames than its receiver can handle.

- **Synchronization**—Ensures that both sender and receiver are capable of establishing and maintaining synchronization.

- **Error detection**—Performs some degree detection of errors.

- **Notification of errors**—Notice of errors.

- **Error correction**—Performs a degree correction of errors in order to implement error recovery.

- **Acknowledgments**—A transmitter must be informed of the correct or incorrect receipt of a frame.

Data Link Protocols

The Data Link Layer is handled by firmware that determines the grouping of bits into frames. Standards have been established to describe how bits should be grouped for the purpose of creating frames; these standards are called data link protocols.

Data link protocols in general describe how frames are carried between systems on a single data link. These include protocols designed to operate over dedicated point-to-point, multipoint, and multiaccess switched services, such as Frame Relay.

Error-Free Communication Paths

The primary function of a data link protocol is to convert an error-prone Physical Layer into an error-free communication path to be used by application processes. If an application could treat the link as though it were error-free and not have to do any error checking in transmitted or received data, it would be much more efficient. Transmission impairments on the telephone lines serve to introduce a non-trivial error rate into telecommunications. The agreement to perform the conversion of an error-prone physical link into an error-free communication path is collectively know as a data link layer protocol.

Using a data link protocol, the two ends of the link agree on a set of procedures to test each incoming message for errors, and to request retransmission of data in error. With these procedures in place, incorrect data theoretically never arrives at the terminating point. Erroneous data frames are discarded and a new copy is retransmitted; no erroneous data ever proceeds from the transmitting process to the receiving process.

Figure 6.2
The Data Link Layer
protocol function.

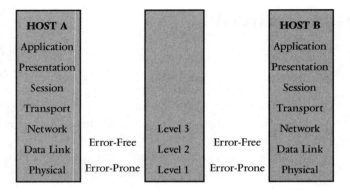

The following functions must be performed if the error-prone physical link is to appear error-free to communicating processes:

- Framing
    - A critical function
    - Receiver is able to determine where a transmitted entity begins and ends
- Error Detection
    - Accomplished by *Cyclic Redundancy Check* (CRC) (explained in further detail later in this chapter)
    - If the receiving station fails to recognize the beginning framing information, no check of a cyclic redundancy remainder will be performed even though an error has occurred
- Sequence Check
    - Transmitting protocol machines append a sequence number with each transmitted data block
    - Loss of block is detected by the arrival of an out-of-sequence transmission
    - Negative acknowledgment is sent out by the receiver
    - Retransmission of the missing block is sent from the transmitter

- Flow control

  — Flow control is essential if link correctness is to be maintained

  — If the receiver's buffer is full, it cannot accept any additional transmitted data

  — Receiver protocol machine will notify the transmitter that further communication is impossible

  — Transmitter protocol machine turns off the flow of information from the parallel section

  — Receiver protocol machine sends out a notice that its buffer is clear

  — Transmitter protocol machine then turns on the flow of information

- Time out

  — If the missing sequence number in a frame is the last number, the receiver protocol machine is not aware that there is a loss of block

  — After a duration of time, the transmitter protocol machine issues a "time out," because it did not receive a positive or negative acknowledgment

  — Receiving protocol machine indicates that it has received the beginning blocks and is awaiting arrival of the final block

  — Transmitting protocol machine then retransmits the last block

Framing

Framing is the first responsibility of any data link protocol. As stated above, framing designates the beginning and the ending of a data block. For example, BISYNC uses special characters to accomplish framing functions: A synchronous block begins with the special character "start of text" and terminates with the special character "end of text."

Another protocol utilized to detect accuracy of framing is called *Digital Data Communication Message Protocol* (DDCMP). This protocol posts in the count field the number of bytes in the text block. There is no requirement for terminating framing information. The receiver knows when it is at the end of a data message when the count has been exhausted.

In protocols that are categorized as bit-oriented protocols, the text is framed with a specific bit pattern (01111110) instead of special characters. A procedure called *zero bit insertion* or *bit stuffing* is used to prevent the flag string from appearing within the text.

If the special characters used for control purposes appear as part of the text, a procedure called *transparency* must be employed. Transparency informs a receiving station that a particular control character should be viewed as text in the given message.

Transparency

Transparency makes a data link protocol invisible to the procedure that initiates the communication. If the communicating process includes a string identical to the flag in the data to be transmitted, the data link protocol machine must temporarily modify the transmitted data to prevent the occurrence of the flag string:

- The data link protocol examines the data passed down for transmission.

- When five consecutive 1s are found in the text, the transparency procedure inserts a sixth 0 following the fifth 1, which is always inserted regardless of the fact that the subsequent bit may itself be a 0.

- The data link protocol machine, upon observing a 0 following a stream of five 1s, removes it and understands that it is not at the end of the transmission.

- If, after a string of five 1s, the receiving station finds a sixth 1, it understands that this must be the flag and the transmission block is at an end.

- The transparency procedure provides a mechanism for preventing premature termination, but it removes any concept of 8-bit multiples in the transmitted data block.

- Because of the presence of stuffed 0s, the transmitted block may well not be a multiple of eight bits in length.

- This protocol is termed *bit-oriented* to distinguish it from protocols that send only multiples of eight-bits, the so-called *byte-oriented protocols*. A bit-oriented protocol uses Cyclic Redundancy Check (CRC) for error detection.

Sequencing

The problem of sequence number failures in the bit-oriented data category is handled by the following procedures:

- A sequence number is included in each transmitted block.

- An acknowledgment number is provided as part of the transmitted frame, which identifies the next frame expected from the other end of the link, rather than specifying the number of the last frame received.

- There is no requirement that every frame be acknowledged.

- The acknowledgment is an implied acknowledgment for all preceding frames.

Time Outs

Time outs are used to prevent deadlock situations from occurring when a reject from a not-received frame is not acknowledged by the receiver. When the sequence of the frames is incorrectly transmitted:

- The time out is issued after an agreed time period.

- The transmitter resends all unacknowledged frames, and transmission resumes as normal.

If the last frame of the sequence is correctly received and the acknowledgment is damaged on its return trip:

- The transmitter will still time out.

- The transmitter retransmits the unacknowledged frame.

- The receiver recognizes the frame as a duplicate (it contains a previously used sequence number).

- The receiver does not accept the incoming frame.

- The receiver acknowledges receipt of the frame to prevent the continued retransmission of the duplicate information.

Flow Control

Flow control monitors the line for congested data traffic. Flow control procedures are defined as follows:

- When a receiver's station buffers are full, further data will not be accepted.

- The receiver sends out notification that no additional data will be accepted.

- The data link protocol machine transmits a supervisory frame called *Receiver Not Ready* (RNR) to the transmitter.

- The transmitter ceases communication.

- The transmitter begins a timer.

- When the timer times out, the transmitter sends the next frame in sequence.

- The receiver responds:

  — If buffer space is still not available, the Receiver Not Ready (RNR) is sent to the transmitter.

  — If buffer space is available, an acknowledgment is sent to notify the transmitter that the not ready condition has terminated prior to receipt of the data frame.

- If the receiver becomes ready before the transmitter timer ends, the receiver sends a supervisory *Receiver Ready* (RR) to signal the transmitting station that it may once again begin sending data.

Data Link Protocol Types

There are two primary types of data link protocols:

- Positive Acknowledgment or Retransmission (PAR)
- Automatic Reply Request (ARQ)

Positive Acknowledgment or Retransmission (PAR)

After transmission of data, a Positive Acknowledgment or Retransmission (PAR) will:

- Wait for an acknowledgment (ACK) reply.

- If a reply is not received before the reply timer expires, the sender will retransmit the frame.

- If the received message has bit errors, the receiver is not required to take any action because the timeout will cause a retransmission—because of this procedure, negative acknowledgments are not needed in this scheme.

Automatic Reply Request (ARQ)

Automatic Reply Request (ARQ) protocols are more common than Positive Acknowledgment or Retransmission (PAR) protocols.

After transmission of a frame, the Automatic Reply Request (ARQ) will:

- Receive either an acknowledgment (ACK) or a negative acknowledgment (NAK) reply.

- If a reply is not received before a reply time out, the sender will send a reply request (REP) message, asking the receiver for the status of the outstanding frame.

There are two types of Acknowledgment or Retransmission (PAR) protocols:

- Stop-and-Wait protocol
- Pipeline protocol
  - Go-Back-N Protocol
  - Selective Retransmission Protocol

STOP-AND-WAIT PROTOCOL

- Only a single message may be outstanding at any given time.
- The transmitter must stop and wait for a reply after every transmission.
- The transmitter and receiver window has a size of 1.

PIPELINE PROTOCOL

- Many messages may be outstanding at the same time.
- The transmitter window size is greater than 1.
- There are two types of pipeline protocols:
  - Go-Back-N
    - More commonly used than Selective Retransmission protocols
    - Utilized in very noisy environments
    - When a frame is negatively acknowledged, the transmitter must retransmit the erroneous frame and all frames that were subsequently transmitted
    - Receiver must receive all frames in sequential order
    - Transmitter window size cannot be greater than the modulo of sequencing
    - Maximum transmitter window size is modulo-1

— Selective Retransmission

- When a frame is negatively acknowledged, the transmitter must retransmit only the erroneous frame

- Receiver may receive frames that are not in sequential order

- Must buffer messages and be able to reorder them to send to the higher layers

- Transmitter window size must be equal to the receiver window

- Each window may not be larger than modulo-2

- Provides a more efficient use of the link than Go-Back-N protocols

- Simpler to implement than Go-Back-N protocols

Implementation of Data Link Protocol

Implementation of a data link protocol can be accomplished in:

- Software

  — Located on a device between the transmitter and receiver

  — Placed on the receiver

  — Not desirable because overhead of processing time

- Hardware

  — Separate processor

  — Internal microprocessor

  — Most common practice is to provide a microprocessor-based solution

Standards for Data Link Protocols

The International Standards Organization (ISO) recommends use of data link protocols. The most commonly used data link protocols in networks are:

- High Level Data Link Control (HDLC)

    — A bit-oriented synchronization protocol that specifies the transmission rules of a signal frame between one device and another over a single data link

    — Specifies a data encapsulation method on synchronous serial links using frame characters and checksums

    — May not be compatible between different vendors

    — Supports both point-to-point and multipoint configurations

    — Subsets of HDLC are:

        - *Synchronous Data Link Control* (SDLC); which is defined by IBM's SNA architecture

        - *LAP and LAP-B*; which is part of CCITT Recommendation X.25

- Frame Relay

    — Utilizes high-quality digital facilities at speeds of T1 (1.544 Mbps) and T3 (44.7 Mbps)

    — Contains no error correction mechanisms, so it can transmit Layer 2 information very quickly

    — An industry-standard, switched data link protocol that handles multiple virtual circuits using HDLC encapsulation between connected devices

    — More efficient than X.25

- Point-to-Point Protocol (PPP)

    — Provides router-to-router and host-to-network connections over synchronous and asynchronous circuits

    — Contains a protocol field to identify the network layer protocol

- Integrated Services Digital Network (ISDN)
    - Set of digital services that transmits voice and data
    - Communication protocol that permits telephone networks to carry data, voice, and other source traffic
- Binary Synchronous Communications (BISYNC)
    - Standard for synchronous terminals
- Digital Data Communications Message Protocol (DDCMP)
    - Data link protocol for DECnet
- Synchronous Data Link Control (SDLC)
    - IBM's SNA
- Local Area Network (LAN) standards for bus or Token Ring networks.

Categories of Data Link Protocol

The two major categories of data link protocols are:

- Binary-synchronous protocols
- Bit-oriented protocols

Binary-Synchronous Protocols

A binary-synchronous protocol is a member of the *character-oriented protocol* group, which was introduced in the 1960s by IBM. The character-oriented protocol family contains the following high-level definitions:

- Contains reserved control functions within a particular code set for transmission
- All frames start with at least two synchronization (SYN) characters

- Special provisions must be made for transmitted messages so that their content contains characters ordinarily reserved for control functions

- Delimits a frame by specified control characters at the beginning and end of the frame

The *Binary-Synchronous Communications* (BSC/BISYNC) protocol contains a set of rules for synchronous transmission of binary coded data. BSC operations encompass:

- A stop-and-wait, automatic reply request protocol

- An industry standard for block-mode (synchronous) terminals

- A half-duplex protocol

- A character-oriented protocol that transmits messages consisting of strings of characters (does not specify length or data code of the characters)

- Bit errors are detected using either:

  — Cyclic Redundancy Check (CRC) polynomials

  — Combination of vertical and longitudinal *parity checking*

- Frames with bit errors prompt a negative acknowledgment reply

- A header is optional; if a header is present, it will be preceded by a Start-of-Header (SOH) character

- The start of the data field is delimited by the Start-of-Text (STX) character

- The end of the Data field is delimited by the:

  — *End-of-Text* (ETX) character

  — *End-of-Transmission Block* (ETB) character

  — *Intermediate Transmission Block* (ITB) character

- The *Block Check Character* (BCC) contains error detection information

- Transparency is provided using a scheme called *character* or *byte stuffing*

- When operating in transparency mode, all special characters (STX, ETX, ETB, etc.) are preceded by a *Data Link Escape* (DLE) character

- Commonly used control characters:

  - Delimit blocks and divide messages into blocks

  - Control half-duplex data exchange

  - Transmit data that might include bit patterns of some control characters handled by the DLE

  - Transmit two SYN characters before sending data

  - Transmit two PAD characters before message to ensure the sending and receiving stations are in bit-synchronization before transmission begins

  - Send two consecutive SYN characters after the transmission of the PAD to establish character synchronization between the sending and receiving stations

  - When transmitting extensive long frames of data, maintain synchronization with the receiving station by inserting two SYN characters into the data stream approximately every second

  - Send a minimum of two PAD characters after the frame is transmitted to end the frame

The BISYNC protocol supports only three data codes:

- ASCII

- EBCDIC

- SBT (6-bit transcode that is rarely used today)

The receiving station verifies that the frame was received correctly by checking the Block Check Character (BCC) sequence. Methods of error detection are based on the BISYNC implementation and character code that is being utilized:

- ASCII

  — Utilizes two forms of error detection:

    - *Vertical Redundancy Check* (VRC)—message text contains a parity bit with each byte

    - *Longitudinal Redundancy Check* (LRC)—1-byte BCC sequence

- EBCDIC

  — Does not use VRC

  — CRC method is used to generate a 2-byte BCC sequence

The transmitting station passes the entire BISYNC frame through an arithmetic algorithm to produce a 1- or 2-byte BCC sequence. LRC or CRC algorithm result values are sent in the BCC sequence with the frame. Frames received are passed through the same algorithm to produce a like value. If values are different, the receiver requests retransmission of the frame.

Figure 6.3
BISYNC protocol.

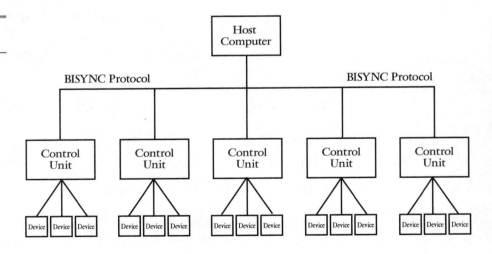

BISYNC does not correct errors—it only produces error detection. The BISYNC protocol is an unbalanced configuration consisting of one master station, referred to as the *host* and several *control units* (CU).

One master host computer can have a maximum of 32 control units connected. Each control unit can have up to 32 devices. The BISYNC protocol is in operation between the host computer and the CU.

Binary-Synchronous Communications (BSC) Protocol "Polling" Service

The BISYNC host operates under the control of a *polling service*. When the host inquires for any information that needs to be sent out, it transmits a data frame containing the control unit address to the polled control unit. Upon receipt, the control unit responds by either transmitting its data or indicating that it does not have any data to be sent. BISYNC is a stop-and-wait protocol, meaning that the host always waits for acknowledgment of receipt for each block of data it sends before it sends another.

Because BISYNC is a half-duplex transmission method, the host computer alternates its transmission state, transmitting to one station at a time. Control units can only transmit data after a poll is directed and the host has given permission to send. The host is unable to transmit again until the control unit has completed its transmission of data.

There are two types of polling protocols for the BYSNC protocol:

- General
- Specific

GENERAL POLL PROTOCOL. A *general poll* is a request from the host to the control unit asking if *any* device on the control unit has any data to transmit. When the EBCDIC character code set is used, the structure of the general poll is:

- The host sends out a DESELECT to inform the previous controller that the transmission session is ending and communication will move to the next controller. DESELECT service consists of the first three characters of any poll:

  — Two SYN characters

  — One End of Transmission (EOT) character

- This is followed by:

  — Two SYN characters from host

  — A Start of Header (SOH) character

  — A Start of Text (STX) character

  — A Write/Write-Erase Command

- Two poll addresses are sent, issuing permission to a control unit to transmit

- The host will only poll attached control unit addresses

- Two double quote characters (" ") are sent, indicating acceptance of the device for the general poll

- If a controller has no data to transmit:

  — Two SYN characters are transmitted from the control unit

  — A No Traffic frame is transmitted from the control unit

  — One ENQ character is transmitted from host

  — A Positive (ACK0) Acknowledgment is sent from the control unit

- If controller has data to transmit:

  — A Read Modify frame is transmitted

  — A Status & Sense (S&S) frame is transmitted if a controller has status information to transmit

  — One ENQ character is transmitted from the host

  — A Positive (ACK0) Acknowledgment is sent from the control unit

SPECIFIC POLL PROTOCOL. A *specific poll* is a request from the host to the control unit asking if a *specific* device attached to the control unit holds any data to transmit. The difference between a general poll and specific poll is that the pair of double quotes (" ") is replaced with a device address. When the EBCDIC character code set is used, the structure of the specific poll frame is:

- The host sends out a DESELECT to inform the previous controller that the transmission session is ending and communication will move to the next controller. A DESELECT service consists of the first three characters of any poll:

  — Two SYN characters

  — One End of Transmission (EOT) character

- Two SYN characters are sent from the host followed by:

  — A Start of Header (SOH) character

  — A Start of Text (STX) character

  — A Write/Write-Erase command

  — Two *polling addresses* identifying the control unit that is being requested to transmit from the host

  — Two *device addresses* identifying the device that is being requested to transmit from the host

- If controller has no data to transmit:

  — Two SYN characters are transmitted from the control unit

  — A No Traffic frame is transmitted

  — One ENQ character is transmitted from the host

  — A Positive (ACK0) Acknowledgment is sent from the control unit

- If controller has data to transmit:

  — A Read Modify frame is transmitted

  — A Status & Sense (S&S) frame is transmitted if a controller has status information to transmit

  — One ENQ character is transmitted from the host

  — A Positive (ACK0) Acknowledgment is sent from the control unit

For the purpose of error checking, it is necessary to send a device address twice to the control unit; repetition of the addresses permits the addressed station to do a match comparison.

STATUS AND SENSE (S&S) FRAME. A Status and Sense (S&S) frame can be used in responding to a general or specific poll. Status information is provided to the host regarding the condition of the devices attached to the control unit (available, unavailable, out of paper, etc.). The structure of the Status and Sense (S&S) frame is:

- The host sends out a DESELECT to inform the previous controller that the transmission session is ending and communication will move to the next controller. A DESELECT service consists of the first three characters of any poll:

  — Two SYN characters

  — One End of Transmission (EOT) character

- This is followed by:

  — Two SYN characters

  — A Start of Header (SOH) character

  — %R characters immediately following the Start of Header (SOH) character to identify Status and Sense (S&S) messages

  — One Start of Text (STX) character

  — A Write/Write-Erase Command

- The following characters indicate the status of the device whose address appears in the device address (DA) character:

  — *Poll address* (PA) characters are transmitted once

  — Device address (DA) characters are transmitted once

  — A Combination of Status & Sense 1 (SS1) characters and Status and Sense 2 (SS2) characters is sent

    - If the SS1 character contains a space and the SS2 character contains an & then the device is "unavailable"

    - If the SS1 character contains a B and the SS2 character contains a space, then the device is "no longer busy"

- Each and every transmission frame must be acknowledged

- An error checking method is used to ensure that transmission frames are not lost by alternating the ACK0 and ACK1 acknowledgments

- — The error checking method is CRC-16 using the EBCDIC code if the BCC is two characters

- — The error checking method is LRC using the ASCII code if the BCC is one character

- The polled control unit has control of the line and can transmit multiple frames

- If controller has no data to transmit:

  - — Two SYN characters are transmitted from the control unit

  - — A No Traffic frame is transmitted from the control unit

  - — One ENQ character is transmitted from host

  - — A Positive (ACK0) Acknowledgment is sent from the control unit

- If controller has data to transmit:

  - — A Read Modify frame is transmitted

  - — A Status and Sense (S&S) frame is transmitted if a controller has status information to transmit

  - — One ENQ character is sent from the host

  - — A Positive (ACK0) Acknowledgment is sent from the control unit

In BISYNC, all frame transmission must be acknowledged positively indicating that the transmission was received correctly and conditions are ready for the next transmission. Alternating positive acknowledgments creates an error detection method to ensure that frames are not lost.

SELECT SEQUENCE.   General or specific polling, by definition, requires the host to ask the control unit if it has any data to send to the host station. A *select sequence* is a request from the host to receive data. To implement a poll and select, each control unit is given two addresses:

- If the first address is used, the frame is a specific poll

- If the second address is used, the frame is a select sequence

For example, when the EBCDIC character code set is used, the structure of the specific poll frame is as follows:

- The host sends out a DESELECT to inform the previous controller that the transmission session is ending and communication will move to the next controller. A DESELECT service consists of the first three characters of any poll:

  — Two SYN characters

  — One End of Transmission (EOT) character

- This is followed by:

  — Two SYN characters

  — A Start of Header (SOH) character

  — A Start of Text (STX) character

- A Write/Write-Erase Command

  — Two selection addresses

  — Two device addresses are transmitted, issuing permission to a control unit to transmit

- The host will only poll to the attached control unit addresses for the specific device

- If controller has no data to transmit:

  — Two SYN characters are transmitted from the control unit

  — A No Traffic frame is transmitted from the control unit

  — One ENQ character is transmitted from the host

  — A Positive (ACK0) Acknowledgment is sent from the control unit

- If controller has data to transmit:

  — A Read Modify frame is transmitted if a controller has data to transmit

  — A Status & Sense (S&S) frame is transmitted if a controller has status information to transmit

  — One ENQ character is sent from the host

> — A Positive (ACK0) Acknowledgment is sent from the control unit

Selection addresses must be different from polling addresses to enable the secondary station to determine what operation is being performed. The control unit address is modified, but not the device address.

WRITE/WRITE-ERASE. After completion of the Select Sequence command from the host and a positive acknowledgment (ACK0) from the control unit, the host transmits a frame containing text. The Write/Write-Erase command is used to clear the control unit's buffer in preparation to receive data. There are eight possible options for the Write/Write-Erase command:

- Write

   — ESC 1 character

- Erased-Write

   — EC 5 character

- Erase All Unprotected

   — ESC ? character

- Copy

   — ESC 7 character

- Read Modified

   — ESC 6 character

- Read All

   — ESC 2 character

- Clear

   — A1

- Pseudo Bid

   — F7

WRITE CONTROL CHARACTER. After the Write/Write-Erase command character, the next character is a *Write Control Character* (WCC). The following list of options can be utilized by selection of the Write Control Character:

- Character Printer Format
    - 00—NL/EM characters honored
    - 01—40—Character print line
    - 10—64—Character print line
    - 11—80—Character print line
- Start Print
    - 0—Do not start printer at completion of write
    - 1—Start printer at completion of write
- Sound KD Alarm
    - 0—Do not sound alarm at completion of write
    - 1—Sound alarm at completion of write
- Restore KD To Local
    - 0—Do not restore to local at completion of write
    - 1—Restore KD to local at completion of write
- Reset Attribute Character to Unmodified
    - 0—Do not reset Attribute Character to unmodified prior to writing data or executing orders
    - 1—Reset Attribute Character to unmodified prior to writing data or executing order

ORDER INFORMATION CHARACTER. Following the Write Control Character (WCC) is the *Order Information Character* (OIC). The Order Information Character sets the buffer address and defines where operations are to begin or continue. Data options immediately follow the Order Information Character:

- Set Buffer Address (SBA)

- Start Field (SF)

- Insert Cursor (IC)

- Program Tab (PT)

- Repeat to Address (RA)

- Erase Unprotected to Address (EUA)

The frame is ended with an End of Text (ETX) and Block Check Character (BCC). The control unit responds with a positive acknowledgment.

NEGATIVE ACKNOWLEDGMENT (NAK) CHARACTER. A *Negative Acknowledgment* (NAK) character usually indicates a bad Block Check Character (BCC) or is sent in response to a poll to refuse a selection of line bid. The following list of instances triggers a host to send a Negative Acknowledgment (NAK) character:

- Receipt of a block containing a parity error (ASCII LRC only)

  — Frame in error is retransmitted

- Receipt of a block having an invalid Block Check Character (BCC) (EBCDIC CRC only)

  — Frame in error is retransmitted

- Receipt of a block terminating in or containing enquiry (ENQ) after a Start of Text (STX) character has been received

  — Refusal to a poll

  — Host transmits a general poll

  — Control responds with a No Traffic response

  — Host transmits a specific poll

  — Control unit responds with a negative response

WAIT-BEFORE-TRANSMIT POSITIVE ACKNOWLEDGMENT (WACK). A *Wait-Before-Transmit Positive Acknowledgment* (WACK) is a positive acknowledgment indicating that the receiver did receive the previous message. It is sent by a receiving station to indicate that it is temporarily not ready to receive.

The sequence that permits a Wait-Before-Transmit Positive Acknowledgment (WACK) to fit into the transmission of data process is:

- The host transmits a specific poll to the control unit

- A No Traffic response is transmitted, indicating the controller is available

- A select sequence is transmitted to the control unit by the host and positively acknowledged by the control unit

- The host transmits an Erase/Write frame

- The Control unit responds with a WACK, indicating that the message has been received correctly, but the control unit is not ready for another message yet

- The host continues to transmit select sequences until a positive acknowledgment is transmitted

- Upon receipt of a positive acknowledgment, the host transmits a Write-Only frame

- The previous Erase/Write was received

- It is not necessary now to perform another Erase command

REVERSE INTERRUPT (RVI). A *Reverse Interrupt* (RVI) is used to permit a station to transmit high-priority data by indicating positive acknowledgment for a quick turnaround of the line. When the RVI is received, the transmitting station continues to send data blocks until its buffer is empty; it then sends an End of Text (EOT) to allow the receiving station to bid for the line.

Here is an example of how a Reverse Interrupt (RVI) is used in the transmission of data process:

- The host transmits a specific poll to a control unit (CU) and device

- The control unit (CU) responds with a status signal

- The control unit (CU) sends a message when the device is no longer busy

- The host transmits a select sequence

- The control unit (CU) responds with a Reverse Interrupt (RVI)

- At some point between the No Traffic response from the control unit (CU) and the select sequence from the host, the device status changes and it becomes unavailable

ARPANet IMP—IMP Protocol

The Department of Defense Advanced Research Projects Agency Network (ARPANet) was designed and packet switching was established for data communications.

The *Interface Message Processor* (IMP) is a node on the ARPANet. The line protocol that controls the line between Interface Message Processors (IMPs) is referred to as the IMP—IMP Protocol. The format of the IMP—IMP Protocol is:

- At least two SYN characters precede each frame

- DLE—STX character pairs delimit the beginning of data

- Data fields can hold up to 125 bytes (1,000 bits) of data

- DLE—ETX character pairs delimit the ending of data

- To avoid confusion when a DLE—ETX appears in the data, DLE stuffing is applied

- A remainder from the CRC-24 calculation appears after the DLE—ETX

- A trailing SYN character ends the procedure

IMP—IMP Physical Link

In IMP—IMP Physical Link operation:

- Transmission is divided into eight logical channels

- Each of these channels implements a Positive Acknowledgment or Retransmission (PAR) protocol

- A message containing bit errors prompts an Acknowledgment (ACK) reply

- A message with bit errors prompts no reply

- The transmitter times out after 125 milliseconds and retransmits the message

- The transmitter uses the lowest numbered logical channel available

- The channel is marked BUSY

- While one channel receives a BUSY and is waiting for an ACK, the other seven channels may be active

- The channel becomes IDLE again when the ACK is received

- Provides a Selective Retransmission protocol with a window size of eight

ARPANet Implementation

In ARPANet implementation:

- The process is not classified as a general protocol

- The Interface Message Processor (IMP) is an intelligent device to provide the logical division of the physical link and the buffering of out-of-sequence messages

Bit-Oriented Protocols

Bit-oriented protocols are most commonly used in telecommunication networks. The major points of the bit-oriented protocols are:

- Independence from any particular code set

- Use of one special character called a FLAG character to mark the beginning and ending of a message

- Uses messages consisting simply of bit streams

- All other combinations of bits are treated as valid data characters

Common Bit-Oriented Protocols

Bit-oriented protocols are more commonly used than character-oriented protocols. A few of the most commonly known bit-oriented protocols used today are:

- Advanced Data Communications Control Procedure (ADCCP)
  - ANS X3.66
  - The U.S. national standard Link Layer protocol
  - Derived from SDLC
  - Environments:
    - Point-to-point or multipoint
    - Balanced or unbalanced
    - Full- or half-duplex
    - Modulo 8 or modulo 128
    - Go-Back-N or selective retransmission
- High-Level Data Link Control (HDLC)
  - ISO 3309, ISO 4335, ISO 7809, and ISO 8888
  - Environments:
    - Point-to-point or multipoint
    - Balanced or unbalanced
    - Full- or half-duplex
    - Modulo 8 or modulo 128
    - Go-Back-N or selective retransmission
- Link Access Procedures (LAP)
  - Original link layer in X.25
  - Environments:
    - Symmetric unbalanced
    - Full-duplex

- Point-to-point
- Modulo 8
- Go-Back-N

■ Link Access Procedures Balanced (LAPB)

— ISO 7776

— Current link layer protocol for X.25

— Environments:

- Balanced
- Full-duplex
- Point-to-point or multipoint
- Modulo 8 or modulo 128
- Go-Back-N

■ Link Access Procedures on the D-channel (LAPD)

— Used on the Integrated Services Digital Network (ISDN) D-channel

— Environments:

- Balanced
- Full-duplex
- Point-to-multipoint
- Modulo 128
- Go-Back-N
- Several logical links multiplexed on a single physical channel

■ Link Access Procedures for Modems (LAPM)

— Error-free modem protocol

— Generated by the CCITT for use on modems

■ Link Access Procedures over Half-Duplex Links (LAPX)

— Level 1.5 protocol

— Implements a half-duplex transmission module for using LAPB over half-duplex channels

- Synchronous Data Link Control (SDLC)

  — Devised by IBM

  — The first bit-oriented protocol

  — Environments:

    - Unbalanced

    - Full-duplex

    - Point-to-point or multidrop

    - Modulo 8 sequencing only

    - Go-Back-N

Format for Bit-Oriented Protocol

All frames in bit-oriented protocols have the following format:

- Frame is delimited by a FLAG bit pattern 01111110
- Address field

  — Identifies the secondary station on the link

  — Differentiates between commands and responses

- Control field

  — Identifies the frame type

  — Carries sequencing information

- Information field

  — Contains data from higher layers

  — Contains any number of bits

- Frame Check Sequence (FCS) field

  — Contains the Cyclic Redundancy Check (CRC) remainder information

- — Uses the CRC-CCITT polynomial
- ▪ Frame is terminated by another FLAG
  - — Bit stuffing is used for transparency if the FLAG bit pattern occurs elsewhere in the frame

Bit-oriented protocols use Go-Back-N, in which:

- ▪ A Receive Ready (RR) reply provides acknowledgment
- ▪ A Reject (REJ) frame is used to indicate an out-of-sequence frame
- ▪ Frames with bit errors are ignored
- ▪ Acknowledgment with the RR and REJ frames is the sequence number of the next expected frame

Frame Types

There are three frame types for bit-oriented protocols:

- ▪ Information frames (I-frames)
  - — Contain data from higher layers
  - — Are sequenced
  - — Carry piggy-backed acknowledgments (RR) and the request for retransmission (RET)
- ▪ Supervisory frames (S-frames)
  - — Control the exchange of I-frames
  - — Carry an acknowledgment number
  - — Provide flow control with the Receive Not Ready (RNR) frame
- ▪ Unnumbered frames (U-frames)
  - — Establish, terminate, and control the status of the link
  - — Have no sequencing associated with them

Logical Stations

Bit-oriented protocols define three types of logical stations:

- Primary station
  - Data flow organization
  - Link-level error recovery
  - Transmitted frames are referred to as "Commands"
- Secondary station
  - Controlled by primary station
  - Transmitted frames are called "Responses"
- Combined station
  - Consists of both the primary and the secondary station features
  - May issue both Commands and Responses

These logical stations are defined by two mode configurations:

- Unbalanced
  - Either point-to-point and/or multipoint operations
  - Consists of one primary and one or more secondary stations
- Balanced
  - Only in point-to-point operations
  - Consists of two balanced stations with each station having equal and complementary responsibility for control of the data link

Data Transfer Modes of Operation

Bit-oriented protocols define two data transfer modes of operation:

- Asynchronous Balanced Mode (ABM)

— Balanced configuration

— Efficient use of full-duplex point-to-point link because of no polling

— Transmission may be initiated by either combined station without receiving permission from the other combined station

- Asynchronous Response Mode (ARM)

— Unbalanced configuration

— Rarely used

— Used when a secondary may need to initiate transmission

— Transmission may be initiated by a secondary without permission of the primary

Multidrop Environment

Many bit-oriented protocols are designed to work in multidrop line environments. The operational steps in a multidrop line environment are:

- Primary station polls secondary station to determine whether the secondary has data to transmit

- Primary station identifies the specific secondary station being polled by its address, which is required as part of the data link protocol

- Primary station indicates that a secondary is allowed to transmit by the expedient of turning on a bit called the *poll-final bit* in the transmitted frame

- The full frame format consists of:

— Initiating flag

— Station address

— Sequence number field containing:

- Sequence number

- Acknowledgment

- Poll-final bit

- Transmitted text

- Cyclic Redundancy Check (CRC) and flag at the end of the transmitted frame

■ Upon receipt of a frame with a poll bit set, a secondary is now allowed to transmit to the primary

A primary station has the capability to transmit to one secondary while receiving from a second. The primary could not poll a second secondary while the first is transmitting on the link. This mode of operation is called *Two-Way Simultaneous* (TWS) communication; it is also referred to as *Full-Full Duplex* (FFDX).

Switching Peer-To-Peer Communication

Data link protocols provide no mechanism for controlling switching function. Peer-to-peer communication requires an additional set of protocols that provide mechanisms for controlling the switching function and for enabling a DTE to establish a nonpermanent connection to a peer processor.

If multiple DTE peers want to communicate with one another, the necessary procedures are:

■ Connection of a DTE to all other DTEs to exchange messages requires a path for each DTE—DTE pair.

■ The connection of all these communication paths is called a *mesh topology*.

■ To determine the number of DTEs to be connected, use the formula $N \times (N-1) \div 2$. This number grows rapidly as N increases.

■ Interconnection of peer DTEs utilizes some form of switching function to:

- Reduce the number of paths

- Permit any DTE to transmit messages to any other DTE

■ Addition of the switching function requires additional protocols.

Level 2.5

The Data Link Layer protocols described so far are single-link protocols (a single protocol controls only a single physical link). These protocol, work at the 1.5 layer to mask incompatible assumptions between level 1 and level 2.

When it is necessary to increase bandwidth or redundancy, more than one physical link is connected to two devices. Because multiple physical links still appear as only a single link to higher levels, a different protocol is needed to provide multilink support between level 2 and level 3. These level 2.5 protocols are defined as follows:

- A layer of protocol that resides between the data link (level 2) protocol and the network (level 3) protocol

- The Network Layer (level 3) sees a single logical link

- Each logical link "sees" the same higher layer above it

- The multilink protocol must be able to reorder data frames that get out of sequence because they are no longer guaranteed to follow the same physical path from level 2 to level 2.5.

The network supplier does not want end users to be aware that there are multiple physical links connecting devices, because the network is not going to let the users choose which physical link they can use. Figure 6.4 illustrates the level 2.5 protocol.

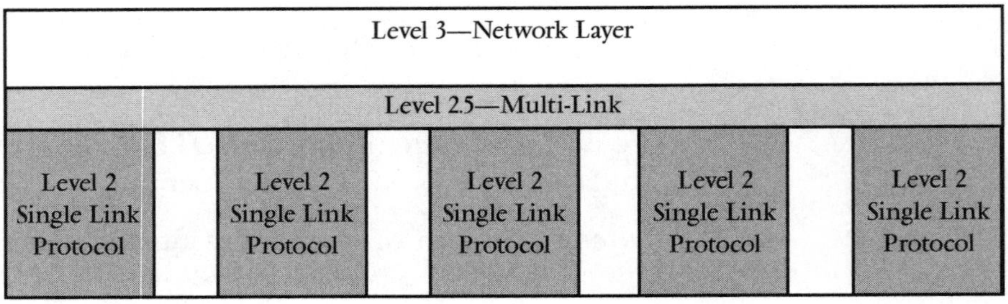

Figure 6.4
Level 2.5 protocols.

Level 3—Network Layer				
Level 2.5—Multi-Link				
Level 2 Single Link Protocol	Level 2 Single Link Protocol	Level 2 Single Link Protocol	Level 2 Single Link Protocol	Level 2 Single Link Protocol

Sliding Window Protocols

A *sliding window protocol* is a protocol that places sequencing numbers in data messages in every data frame (nondata frames are not sequenced). There are three sliding window protocols that work together:

- Transmitter Window Protocol

  — Contains the sequence number of all transmitted frames

  — These transmitted frames are outstanding and have not yet been acknowledged

- Transmitter Window Size Protocol

  — The number of frames contained in the Transmitter Window

- Receiver Window Protocol

  — Receiver maintains a list of legal sequence numbers that it can receive

  — Most common size of the Receiver Window Protocol is 1, which means that frames must be received in sequential order

Synchronization and Asynchronous Functions in the Data Link Layer

Determining which bits belong to which character can be a challenge for the receiving end of a transmission line. The main key is to determine which bit is the beginning bit of a character, how many bits are contained in a character, and the transmission speed of the incoming bits. Two common techniques utilized to determine the first bit of an incoming character are synchronous transmission testing and asynchronous transmission testing.

Synchronous Transmission Testing

In the synchronous transmission testing process, character bits are transmitted without start and stop bits so that the receiver is only required to identify the first bit of the first character. Transmission time is easily determined by the synchronization of clocks between the transmitter and the receiver. Clock synchronization is achieved by:

- Embedding this information in the data signal

- Establishing a clock line between transmitter and receiver

- Determining the frequency of the carrier phase to the receiver, when analog signals are transmitted

- Using biphase encoding, when digital signals are transmitted

When determining the transmission speed and size of character, the receiver can count off groups of bits and correctly assemble the incoming characters. In the case of ASCII (which is an 8-bit numbering system), it is a simple matter of counting off groups of eight bits.

Each block of data starts with a bit pattern and ends with a bit pattern. Together, the control information and the data make up a frame that can be formatted either as:

- Character-oriented, where each block of data begins with two SYN (odd parity) transmission control characters at the beginning of the data stream.

- Bit-oriented, where each block of data begins with a FLAG that is expressed in the binary value of 01111110 (hex value 7E), control fields, data field, more control fields, and repeated FLAG.

Synchronous transmission testing:

- Uses a variety of interfaces, such as RS-232, X.21, V.35, etc.

- Requires timing leads

- Transmits whole blocks of data instead of one character at a time

During a synchronization connection, the Data Link Layer protocol performs a number of tasks to ensure an error-free data link. These tasks include:

- Synchronization between two stations

  — Placement of bits, bytes (characters), and frames

  — Without synchronization, lost or duplicate frames cannot be detected

- Sequence numbers

  — Essential to accomplish synchronization

- Framing

  — Indicates the beginning and ending of a frame

- Delimiting data

  — Specifies the location of the data

- Error control

  — Detects and repairs bit errors or out-of-sequence frames

- Transmit control

  — Informs stations when they are allowed to transmit their data

- Flow control

  — Prevents a transmitter from sending more information than a receiver can accept

- Transparency

  — Framing information is carried as data and not misinterpreted

- Addressing

  — Station that is transmitting or receiving data can be identified

- Link initiation and termination

  — Establishes and terminates the link

To deal with errors on a synchronous link, three approaches are utilized:

- Forward Error Correction (FEC)

- Detect and Retransmit—Go-Back-N

- Detect and Retransmit—Cyclic Redundancy Check (CRC)

FORWARD ERROR CORRECTION (FEC). The Forward Error Correction (FEC) scheme for synchronous connections:

- Does not require a station to wait for an acknowledgment of a block before transmitting additional data

- Transmits overhead with data to enable the receiving station to detect the presence of an error

- Uses overhead to point out offending bits

- Has a high ratio of offending bits to data bits for error detection—if the high ratio of bits are incorrect, most likely the data bits are also incorrect

- Is practical in situations where there is a long delay, which is characteristic of a communication channel

- Works well on noisy lines

- Contains no reverse channel

DETECT AND RETRANSMIT—GO-BACK-N. Detect and Retransmit—Go-Back-N connections:

- Are more commonly used than Forward Error Correction

- Do not require a station to wait for acknowledgment of a block before transmitting additional data

- Use a "stop and wait" procedure that causes the transmitter to sit in an idle condition for almost a half a second before receiving an acknowledgment for a transmitted block

- Allow a transmitting station to "pipeline," thus enabling a transmitter to have many blocks outstanding without acknowledgment

- Upon receipt of a bad transmission, the receiver sends a negative acknowledgment and discards all subsequent frames:

  — Transmitter retransmits the offending frame

— All subsequent frames are discarded by the receiver

— After retransmission, the receiver obtains all the data blocks correctly and in correct sequence

DETECT AND RETRANSMIT—CYCLIC REDUNDANCY CHECK (CRC). The Cyclic Redundancy Check (CRC) will detect an error in the transmitted data block. This mechanism is used by all synchronous line protocols. The steps for this error detection mechanism are:

- A generator, which is a given sequence of ones and zeros is used. The transmitter and the receiver have generator information stored in their read-only memory

- Transmitter treats the message as a number

- Transmitter divides the message by the generator string, which leads to both a quotient and a remainder

- The quotient size is as large as the string, and the remainder is smaller than the generator string

- The generator string will be the remainder, which is sent to the other end of the line as the error-checking string

- The remainder is appended at the end of the transmitted message

- The receiving stations:

  — Divide the incoming message by the generator with the same calculation that was performed by the transmitter

  — Determine a remainder

  — Match the remainders

    - If the remainders do not match, an error must have occurred over the transmission facility

    - The receiving station notifies the transmitter that it must resend the offending block

Cyclic Redundancy Checking (CRC) has a very low probability of failure, less than 1/100 of 1 percent. Cyclic Redundancy Checking

(CRC) adds no time overhead to the transmission because the division is performed in parallel with the actual transmission of the message.

Asynchronous Transmission Testing

In the asynchronous transmission testing process, time management is unnecessary because characters are transmitted one at a time. An example is a terminal that does not contain any buffers to hold data—data is transmitted to the terminal screen as it is typed onto the keyboard.

A receiver recognizes the first bit of each character by the use of start pulses which are detected by monitoring the condition of the line. The two condition states of a line are:

- **Idle**—The transmitter sends a string of 1s, or a *stop bit.*

- **Transmission of character**—Any character beginning with a 0, also referred as *start bit.*

Asynchronous transmission testing is often called *start-and-stop bit transmission testing.*

Steps in the asynchronous transmission testing procedure are:

- Upon detection of a 1 state, the receiver begins clocking

- The clock notifies the receiver, only half a bit length later, to verify if the line exists in the 0 state

- If so, receiver accepts 0 as a start bit

- The state of the line is checked at intervals of one bit length

- Incoming characters are assembled

- Clock is resynchronized at the beginning of each character

Asynchronous Transmission:

- Commonly uses RS-232 interface

- Adds start (0) and stop (1) bits to each character

- Transmits one character at a time

- Uses a parity bit

- Determines length of character depending on code used

During an asynchronous connection, the Data Link Layer protocol uses *echoplexing* of incoming characters to ensure an error-free data link. Echoplexing requires manually checking output to determine if it does not match information that was input.

Asynchronous transmission testing requires 25 percent more overhead than synchronous transmission because of added control information. Therefore, synchronous transmission testing is more efficient for transmitting large blocks of data.

Network Layer
(Level 3)

The Network Layer (Figure 7.1) is Layer 3 of the OSI Reference Model. This layer is responsible for establishing, maintaining, and terminating the network connection between two transport entities and for transferring data to higher layers of the OSI Reference Model.

Figure 7.1
Network Layer in the OSI Telecommunications Reference model.

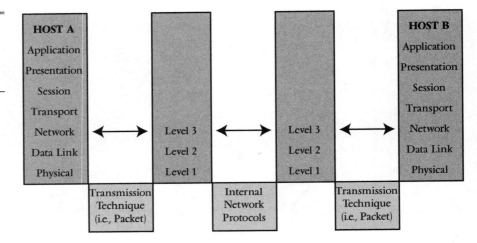

There can only be one network connection between two nodes. However, there can be multiple possible physical routes between two transport entities. The Network Layer handles the routing path between telecommunication machines, sometimes using a number of physical communications links. Each physical link spans two network machines, which must use (at least) Physical and Data Link Layer procedures to exchange data. A network machine may require that messages be divided into pieces called *packets*. To reassemble these packets, the Network Layer creates a *virtual circuit* environment that delivers the packets to the original transmitter in the same sequence as they were originally sent.

The Network Layer is responsible for controlling the passage of packets from their source to their destination in the network. It controls these passages by attaching its own control information to the packets. The addition of these bit controls directs data transmission to be sent over physical paths (i.e., cable), logical paths (i.e., software), and network routers. *Routers* are physical devices that contain software that enables packets formatted on one network to reach a different network in a format that the second network understands. By controlling the passage of packets, the Network Layer acts like a switching station, routing packets along the most efficient of several different paths.

The Network Layer creates logical communication paths that are set up to send and receive data. These paths are referred to as virtual circuits, and are known only to the Network Layer. The network layer is responsible for managing and tracking data from various virtual circuits. If the data packets are not in proper order, the Network Layer resequences the data for proper legibility in level 4.

Other functions of the Network Layer are to:

- Address packets

- Ensure and maintain ordering of packets sent through the network

- Issue acknowledgment receipt of an entire message

- In a connection-oriented network, establish logical connection between two end systems before the system can exchange packets

- Break down transport (level 4) messages into blocks for proper transmission size

- Determine how packets route from one end system to another end system:

  — Virtual circuit networks, route one time per call

  — Datagram networks, routing is done for every packet

- Resize packets to the sizing requirements of the receiving network

- Ensure that subnetworks are not swamped with too many packets from all of the end systems

- Ensure that packets are not sent at a higher speed than the receiving layer can handle

- Ensure that network monitoring, traffic pattern analysis, error analysis, and packet transfer rate calculation are performed

To summarize transmission through the OSI Reference Model so far:

- Bits travel over the Physical Layer

- The Data Link Layer ensures that the physical transmission is error-free

- The Network Layer determines where the bits are sent:

  — If destination is to another machine:

    - Bits go back down to the Data Link and Physical Layers

    - Physical Layer connects to the other machine's Physical Layer

Thus, in telecommunications, a node never implements high layers because they are never invoked.

Switching Network

Recall that we previously discussed that the number of links required to provide interconnection for a switching network is $N \times (N-1)/2$. Using this formula, the process of providing interconnection among DTEs is unnecessarily wasteful of facilities. To maintain a reasonable number of facilities, a switching network is introduced to provide links between DTEs on a nonpermanent basis. See Figure 7.2.

Figure 7.2
Switching network.

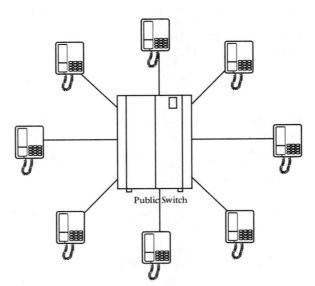

Public Switch

Voice and Data Bandwidth Requirements

Bandwidth is the available transmission space or unused capacity of a network needed the transmission of data or voice.

- Data transmissions
    - Utilize large bandwidth
    - Computer machine processing data speeds are greater than the normal communication transmission or bandwidths
    - Require large quantities of bandwidth only for a brief time interval; this is known as *bursty* transmission
    - Have limitations to the time delay that can be injected into a transmission
    - Are delay-insensitive under a reasonable amount of delays
- Voice transmissions
    - Require little bandwidth
    - Hold the bandwidth for a long period of time with high utilization
    - Have a high degree of delay sensitivity
    - It is critical that a communication facility never insert delays on its own into voice transmission
    - If delays are inserted, a voice transmission cannot be properly interpreted on the other end

There are unusual situations when voice and data transmissions do not perform according to the general rules—for example, data communications may exhibit a long hold time and voice communications may be very brief.

Because of their fundamental differences, it is desirable to transmit these two signals over fundamentally different network structures, based on the different responses of voice and data to system-induced delay.

Circuit Switching

The term *circuit switching* describes a form of telephone transmission switching that provides connection between communicating parties over dedicated facilities. This communicating mechanism connects parties by establishing a dedicated path between the two parties. The telephone network has shared facilities to provide voice communication to many different connections.

By using *frequency division multiplexing,* the conversation may own a frequency band (time slot). No one shares the specific band on which a given telephone conversation is being carried; a telephone conversation proceeds over the equivalent of a dedicated pair of copper wires.

For circuit switching to properly function, *multiplexing* is used to provide dedicated resources and the sharing of physical facilities. Otherwise, an additional line would be required for transmission. Multiplexing creates multiple narrow band paths, which work ideally with voice transmission.

In multiplexing facilities, each conversation establishes a point-to-point connection for the duration of the call by owning its own dedicated resource. If a voice signal were required to share a path with another signal, the attendant delay in waiting for a pause in the other conversation would lead to system-induced delay in the transmission of the voice signal.

Voice and Circuit Switching

Voice is delay-sensitive and has narrow bandwidth (300—3300 Hz) requirements. Voice also requires a dedicated path, thus it functions well in a traditional switched and multiplexed telecommunication network. When a call is initiated, time may elapse before a dedicated connection is established because the circuit-switching network must find an unused end-to-end path.

In the United States, telephone company tariffs are written to charge more for the connection phase of a call than for the ongoing conversation phase. This allocation of charges is based on the period when a call is searching for an unused end-to-end path and expensive telephone company equipment is being utilized. In voice communication, the long hold time of a voice signal and the high utilization of network facilities make the high cost of establishing a connection to be expected.

Data and Circuit Switching Incompatibility

Data transmissions are inconsistent in a circuit-switching environment because:

- Data requires wide bandwidth; circuit switching provides narrowband transmission

- If a transmission allocates 100 KHz to each data communication, the number of physical facilities required will be 33 times as great as the number of facilities required for voice communication

- Data is bursty and requires fast setup; circuit switching requires long setup times. In a circuit-switching environment, this setup leads to high cost and wasted facilities

- Overhead in establishing a circuit-switched connection is extremely large and uneconomical

- The delay-insensitivity of data does not require the dedicated facilities associated with circuit-switching

DATA AND CIRCUIT SWITCHING EXCEPTIONS. A few specific areas of application technology require the circuit-switching environment. Two types of data applications that may require the dedicated facilities associated with a circuit-switched connection are:

- Nonbursty

  — Transfer of large files requires a long hold-time communication

- Real-time transmission that requires the guaranteed delay property of a circuit connection

  — Real-time communications require the receiver's response to occur within the transmitter's time frame. This type of transaction cannot live with the delay associated in alternative forms of switching

Circuit-Switched Summary

The circuit-switched network:

- Provides a dedicated path between communicating parties
- Maintains a dedicated path for the duration of a connection
- Provides a circuit as a fixed bandwidth entity
- Provides fixed delay within each circuit
- Never injects an intersignal delay

Nonswitched Dedicated Path—Private Line Networks

A dedicated path on a nonswitched network is desirable when transmitting a high volume of traffic between systems. The advantages are:

- Path is always available
- No setup is required
- Large number of traffic bursts is exchanged on a regular basis, thus causing less overhead for continual connection setup
- Provides higher data rates than dialup lines
- Full bandwidth is available in both directions
- Less noisy than switched lines

A few private line drawbacks are:

- High cost to maintain a dedicated line
- Charge is appropriated to user even when line is not in use
- Must have significant traffic volume to be cost effective

Statistical Multiplexers (statmux)

A *statistical multiplexer* (statmux) connects devices located in various cluster locations to devices at several other cluster locations. A statmux is located at each of the cluster locations, and the DTEs are connected to the statmux over dedicated facilities. Cost is nominal because the statmux is close to its serving DTEs. If a device must send data to another device, the data first is transmitted to the statmux with appropriate addressing information. If the link is currently in use, the data is buffered until the link becomes available.

The statistical multiplexer functions well with bursty data transmission, because stations are usually available whenever a given station is ready to transmit. If the link is not immediately available, there may be a brief delay. But because data is delay-insensitive, it can live quite comfortably in this environment.

Statmuxes are synchronous devices. Transmission from statmux to statmux must carry an address. Because the statmux appends the address to the entire block of data instead of appending the address to each character, it saves a large burden of overhead. If the DTEs are asynchronous, the link between statmuxes is synchronous. Statistical multiplexers buffer character streams until a full data block is ready for transmission. This functionality is often called *packet assembly and disassembly* (PAD), and is a common requirement for statistical multiplexing equipment. Figure 7.3 shows a statistical multiplexer.

Figure 7.3
Statistical multiplexer.

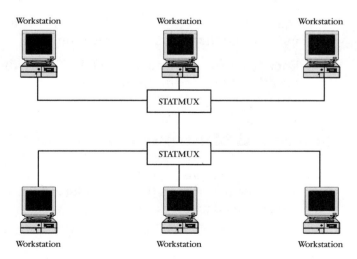

The statmux has the capability to control several output links through the use of a table of destination addresses. When a DTE submits a block for transmission, the statmux selects the appropriate link for transmission, buffering the data until the line in question becomes available (Figure 7.4).

Figure 7.4
Multiplexing multiple
output links.

DTE	R	Buffer for Line 1	Line 1
DTE	O	Buffer for Line 2	Line 2
DTE	U	Buffer for Line 3	Line 3
DTE	T	Buffer for Line 4	Line 4
DTE	I	Buffer for Line 5	Line 5
DTE	N	Buffer for Line 6	Line 6
DTE	G	Buffer for Line 7	Line 7

Some important clarifiers when dealing with statistical multiplexers are:

- Statistical multiplexers that handle multiple lines are used as a network switch that allows communication from one DTE to a collection of other DTEs

- Statistical multiplexer switching function is not circuit-switching, because there are no dedicated links

- Data from one DTE to another DTE will always utilize shared facilities at some point in the path

Statistical Multiplexing Overhead

In a circuit mode transfer, data may be sent with no addressing overhead. Only the destination or source addresses must be included with each transmitted data block.

In the event of bursty data, addressing overhead is a low cost for savings obtained by dramatically reducing the number of dedicated facilities required.

Figure 7.5
Circuit mode versus statmux transmission.

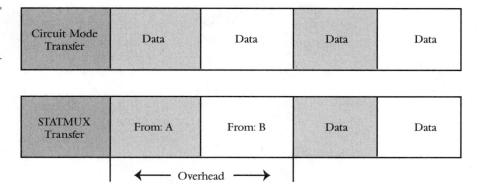

Store-and-Forward Network
========================

When multiple statistical multiplexers (statmux) are interconnected, a rich network structure is created, with multiple paths available from any DTE to any other DTE because routing is complex, the nodal processor must select the optimal path. For proper optimal routing there must be protocols among the nodes that enable them to convey delay information to one another. If a node wants to route a transmission over the best path, it will forward it along a path for which other nodes have reported light delays. This process is known as store-and-forward; the network is shown in Figure 7.6.

The transmission steps for store-and-forward are:

- DTE drops a data block into its attached node

- Node performs a routing calculation

- Data is queued for the selected link

- Upon availability of the link, the data is forwarded to the next node

- An identical operation takes place at the next node

Figure 7.6
A store-and-forward network.

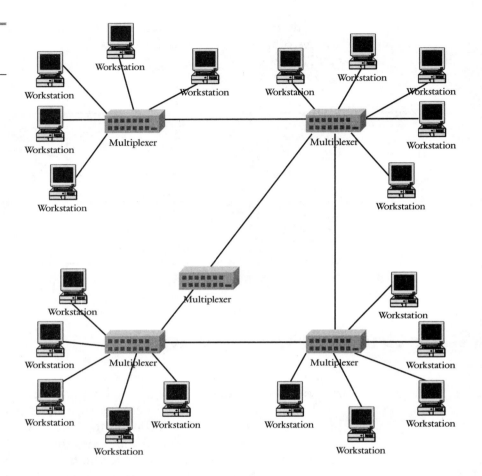

Telegraph Network

The *telegraph network* is one of many varieties of store-and-forward networks. The operational steps in a telegraph network are:

- Telegram is entered on a teletype machine from a local office
- The telegram is transmitted to a *torn tape center*
- The torn tape center is located in a major city and serves many local offices
- Teletypes in both local and remote offices are of a variety known as *Auto Send Receive* (ASR)

- ASR teletypes generate paper tape upon receipt of a transmission from the line
- An operator (who used to work on roller skates, by the way) tears off the incoming telegram
- The operator looks up the destination in a book of routes
- If the destination requires several routes, the operator selects that which is believed to have minimum delay
- The operator proceeds (on skates) to the teletype to transmit the telegram to the next destination
- Telegram is placed in a buffer waiting for the availability of the transmission link
- The buffer resembles the carousel used to hold orders in a luncheonette
- The telegram resides in the buffer or tape holder until it comes to the head of the queue
- The telegram is read through the paper tape reader on the teletype
- The telegram is then forwarded to the next station

Table 7.1 illustrates the relationship between a telegraph network and a modern store-and-forward data network.

TABLE 7.1

Comparison of Telegraph and Store-and-Forward Networks

Torn Tape Center	Store-and-Forward Switch
Local office	Data terminal equipment
Torn tape center	Node
Routing table	Routing table
Buffer tape holder	Memory buffer

Message Switching

The telegraph network is an example of message switching technology. The key characteristics of a message-switched network are:

- Store-and-forward processes are used

- Message switches retain a record of every message long after the message has been sent on its way

- No upper limit on the length of a transmitted message

- A very long message causes delay for other messages waiting in the queue to be transmitted

- Highly variable delay

Packet Switching Networks

Packet switching is an alternative form of store-and-forward switching. The key characteristics of a message-switched network are:

- Store-and-forward network processes are used

- Fixed upper limit on the length of a transmitted entity

- A long message is broken down into short bursts to allow other traffic to be transmitted over the communication facilities without long delays

- Intended to be used principally by interactive traffic

- Buffers are kept in main memory to minimize the delay in retrieving a packet from the buffer and placing it on the communications link

- No archiving is performed

- When a packet has been forwarded, its place in memory is made available for an incoming packet

Packet switching is an improvement on message switching in several ways:

- Queuing delay is smaller

- Ability to intersperse traffic from many sources over a single link is present

- Upper limit on length of transmitted entity is present

- Results in higher performance communication link

- Packet is accumulated and error checking takes place before it can be forwarded

- During packet transmission, another packet can arrive at the switch

- This ability to overlap transmission and reception leads to a lower store-and-forward *queuing delay*

There are disadvantages in sending many large messages in a packet-switching network. It may not be the most cost-effective means of transmission if a user regularly sends messages consisting of several pages of text. Below is a list of disadvantages:

- Every packet carries the same overhead as the entire message would in a message-switched network

- Long messages incur large delay variability

- This problem is particularly acute for bulk data transfer: circuit switching is a better alternative for this type of transmission

- If there is significant amount of other traffic, the message may be substantially delayed by the interjection of traffic from other sources

- If there is no other traffic on the network, the message will propagate rapidly to the destination

- Not an optimal technology for the transmission of large blocks of data

A packet-switching network is optimal for transmission of small data blocks that are one packet long. It also works well for the transmission of very bursty data, where messages are sent with the benefits of packet switching with no increase in overhead when compared to message switching.

Central Office Local Area Network (CO LAN)

A *Central Office Local Area Network* (CO LAN) is a central office—based switched data communications network service that connects terminals and computers using packet switching technology. It uses existing local twisted pair facilities with transmission rates of up to 19.2 Kbps asynchronous and 56 Kbps synchronous. CO LAN is an important service offering, although less frequently utilized nowadays; it meets the needs of business customers who want the features of a local area network (LAN), but cannot purchase their own on-premise LAN. A CO LAN also addresses issues of service integration and efficient data transport.

Premises LANs versus CO-Based LANs

A Premise LAN differs from a CO LAN in:

- Line speed
    - If devices are directly attached to an on-site LAN, the devices operate at the speed of the LAN
    - In a CO LAN, device speed is limited to the weakest link
    - CO LANs limit speed to whatever can be achieved over the local loop
    - Coax or fiber may be utilized if the customer is close to the CO
- Cable distribution
    - CO LAN cable distribution to different sites on the same floor is simple
    - Distribution to different sites on different floors is difficult and more expensive
    - CO LAN costs are minimal
    - CO LAN control is minimal
    - Premise LANs require personnel to operate and configure the network
    - CO LAN network management is off-site
    - Premises LANs depend on media and geographic scope

Channel Allocation

Channel allocation on a CO LAN includes:

- Star networks that utilize any point-to-point medium (i.e., twisted-pair, coaxial cable, or optical fiber)

- Ring networks that may utilize any point-to-point medium

- Bus networks that utilize a medium to support multipoint links: this limits the user to twisted-pair or coaxial cable

- Trees that utilize broadband coaxial cable

Media Usage

Media usage on a CO LAN includes:

- Twisted-pair, which may be used in any point-to-point or multipoint environment—works well on a ring, bus, or star topology

- Twisted-pair may not be used in trees, because trees require transmission in two directions simultaneously

- Baseband coaxial cable may be used in the same environment as twisted-pair

- Broadband coaxial cable may be used:

  — In point-to-point or multipoint environments

  — Commonly only in trees and buses

  — Rings and stars are not practical because the stations in these topologies do not modulate signals—cable and installation is expensive.

- Optical fiber is only appropriate in point-to-point environments; fiber is only used with rings and stars.

Table 7.2 illustrates CO LAN topology and media usage.

TABLE 7.2

CO LAN Topology
and Media Usage

Medium	Ring	Bus	Star	Tree
Twisted pair	✔	✔	✔	
Baseband coax	✔	✔	✔	
Broadband		✔		✔
Coax				
Optical fiber	✔		✔	

All signals leaving and entering the computer are in digital format. A direct computer-to-computer communications pathway is a simple connection between a Data Terminal Equipment (DTE) to an adjacent Data Circuit Equipment (DCE) interface, as shown in Figure 7.7.

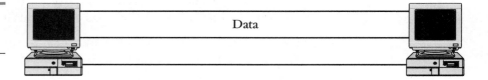

Figure 7.7
Direct data
transmission.

Data

Analog Data Circuits

If the signal between telecommunications equipment is not in digital form, it is necessary to convert the signal to one consistent with the requirements of the receiving telecommunications facility. A typical analog data circuit begins as a digital signal from the customer's equipment. The two components of this analog data circuit are:

- Modem
- *Network Channel Terminal Equipment* (NCTE)

Modems

A modem is an analog-to-digital and digital-to-analog signal converter that works by modulating a signal onto a carrier wave at the originating end and demodulating it at the receiving end. The word *modem* is derived from:

- Modulation
  - Conversion of a customer's digital telecommunication signal into an analog signal (or analog to digital signal) for transmission over a telecommunications network
- Demodulation
  - Conversion of the customer's telecommunication signal back to its original pre-modulation format

The signaling components that are transmitted through the telecommunication lines are described in detail in Chapter 5. In this chapter, we are concerned with the hardware components and layout required for analog data transmission through a modem.

The transmission process through a modem includes:

- Digital output from the computer driver is sent to the local modem
- The modem converts digital signals into analog signal waves to be transmitted over the network
- Analog signals reach the telecommunications facility, where they are converted to digital signals again
- Signals are transmitted over a carrier system
- The modem at the receiver converts the analog signal waves to digital signals consistent with the receiving computer; the whole procedure is reversed at the other end of the circuit

Network Channel Terminal Equipment (NCTE)

Network Channel Terminal Equipment (NCTE) is used to condition the signal. The NCTE provides the following functions in the transmission and reception of telecommunications lines:

- Gain or pad

- Impedance matching

- Equalization

- Loop-back of a signal on demand:

  - Accomplished by sending 2,713 HZ for 5 seconds and then removing it

  - Removes the customer's modem from the transmission loop

  - Additional signals are amplified and returned on the other pair

  - Signal is removed by sending 2,713 HZ again

- Connection of the simplex leads from both pairs (allows for sealing current):

  - −48 VDC is simplexed onto one pair while ground is simplexed onto the other pair in the telecommunication's facility

  - The low voltage flows through the simplex connection in the NCTE and back to ground

  - This current is usually adequate to keep oxidation from forming at connection points in the cable pairs

Figure 7.8 illustrates communication over an analog or broadband line.

Figure 7.8
Communication over
an analog or
broadband line.

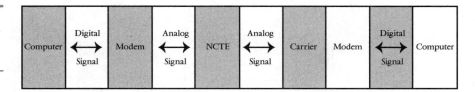

Note: Point-to-point analog data circuits do not use line signaling.

Analog Data Topologies

The most common analog data topologies for telecommunications are:

- Dial-up
- Two-Point
- Multipoint
- Bridge

DIAL-UP. A *dial-up circuit* is a nondesigned dial tone circuit used to access other telecommunications switched data circuits, as in personal computers (PCs). See Figure 7.9.

Figure 7.9
Dial-up data
topology.

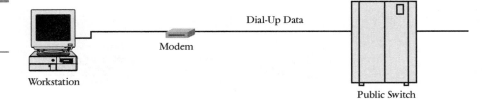

TWO-POINT CIRCUITS. A *two-point circuit* is a dedicated private line data service between two customer premises or a customer premise and an Internet connection. See Figure 7.10.

Figure 7.10
Two-point circuits.

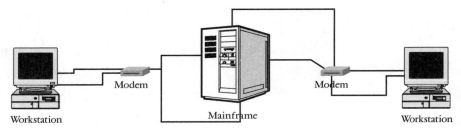

Two-Point Private Line Data

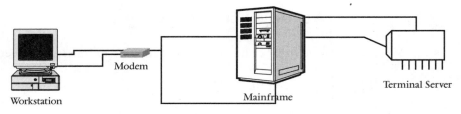

Two-Point Data Access

These connections may extend through one or more Central Offices.

MULTIPOINT CIRCUITS AND BRIDGES. A multipoint circuit is a private line data service connecting three or more customer locations. A bridge(s) must be provided in one or more telecommunications facilities to provide this type of service. See Figure 7.11.

Figure 7.11
Multipoint private data line.

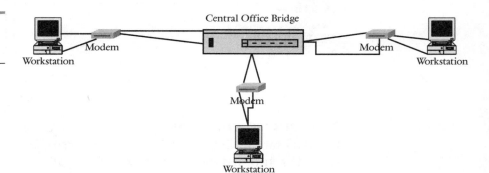

The most common bridge configuration is the *split-bridge* operation, which:

- Uses a separate transmit and receive bridge
- Is wired so that there is a single master location
- Communicates with multiple remote locations

Hardware Communication Testing

It is essential to periodically test communication between hardware devices. Data service tests are measurements used for benchmarks and trouble locations:

- End-to-end measurements
    - Connect customer-interface-to-customer-interface on a two-point circuit
    - Connect customer interface to a bridge
    - Connect customer interface to an end link
- Straightaway test
    - Direct between telecommunications facilities
    - Tests both directions simultaneously between telecommunications facilities
    - Connects a telecommunications facility and the customer location
    - Connects a telecommunications facility and two customer locations
- Loopback test
    - Connects a telecommunications facility to a loopback device and then back to the telecommunication facility

Benchmark measurements are made immediately following installation. *Circuit order tests* are used as a reference for future test purposes. Two common locations to use as benchmark areas for testing are:

- End link

  — The facility between a bridge in a telecommunications facility and the customer's interface.

- Middle link or mid-link

  — Middle link—Equipment between two bridges in the same telecommunications facility

  — Mid-link—Equipment between two bridges in different telecommunications facilities

Conditioning

Conditioning levels determine the proper hardware performance benchmarks required for different transmission objectives:

- C-Conditioning

  — Parameter measurements of attenuation distortion tests

  — Parameter measurements of delay distortion test

- D-Conditioning

  — Parameters of intermodulation distortion tests

  — Parameters of C-notched noise expressed as signal-to-C-notch noise ratio

- E1-Conditioning

  — Optional—Determination based on value limits of the grade of service for message trunk and special service circuit transmission channels that the customer has ordered

- E2-Conditioning

  — Optional—Determination based on value limits of the grade of service for message trunk and special service circuit transmission channels that the customer has ordered

- M1-Conditioning

  — Optional—Determination based on value limits of the grade of service for message trunk and special service circuit transmission channels that the customer has ordered

■ M2-Conditioning

— Optional—Determination based on value limits of the grade of service for message trunk and special service circuit transmission channels that the customer has ordered

C-CONDITIONING.   Table 7.3 lists C-Conditioning designations.

TABLE 7.3

C-conditioning
Designations

C-conditioning Designators	Description
	Basic, 3001, 3002, 3003 VB
C1	Lowest grade for data circuit
C2	Intermediate grade for data
C3	Switched service network only
C4	High intermediate grade for data
C5	High grade for data
C6	Protective relay channels only
C7	Switched services with switch
C8	Customer's location

D-CONDITIONING.   D-Conditioning measures high performance data. It may be required in addition to C-Conditioning or can stand alone with basic conditioning. There are two basic topologies in determining use of D-Conditioning:

■ D1

— A two-point service with no switching and not more than one station per service point

■ D2

— A two- or three-point service where there are no more than three stations per channel

OPTIONAL CONDITIONING SPECIFICATIONS

- E1-Conditioning
  - One end link
- E2-Conditioning
  - Two end links
- M1-Conditioning
  - One mid link
- M2-Conditioning
  - Two mid links

Test Details (TD) Document

The Test Details (TD) document details the required tests and the test parameters. Contents of the Test Details (TD) include:

- Required tests
  - Itemized list of all required tests
- Preservice test limits
  - Required when the circuit is installed or rearranged
- Tariff/Immediate Action Limit (IAL) test limits
  - Tests are guaranteed to the customer and immediate corrective action is required
- Maintenance Test Limits
  - Performed to isolate trouble on a circuit after it has been turned up or handed off to the customer

Digital Transmission

The Physical Layer hardware components required for digital data transmission are:

- In digital data transmission, output from the transmitter must compare the interface type and characteristic at the receiver

- A *Channel Service Unit/Data Service Unit* (CSU/DSU), located between the computer and the digital communications facility, to analyze the digital signals

- A *Data Service Unit* (DSU), to perform the signaling conversion

- A *Channel Service Unit* (CSU), to perform electrical correctness tests over the communication path

Figure 7.12 illustrates communication over a digital line.

Figure 7.12
Communication on a digital line.

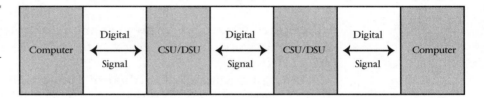

Channel Service Unit (CSU)

The Channel Service Unit (CSU) regenerates "like" signals and provides protection for the network. Some of the functions the CSU offers are:

- Testing capabilities

- Maintaining clear data channels by inserting special codes

- Monitoring the data stream

- Generating a "keep-alive" signal to maintain network timing

- Accepting clocking from either the host or the network

Data Service Unit (DSU)

The Data Service Unit (DSU) regenerates "unlike" signals. DSUs are used when the host is providing serial data in some format other than that used by the receiver. DSUs can also offer a secondary channel for testing purposes, which will include diagnostic capability.

CSU/DSU Combination

CSU and DSU functions are combined in a single unit when transmission configurations are common and speeds are low. Higher bit rates, such as 1.544 Mbps and T1 signals, use only the CSU conversion because normally they do not require the signal conversion.

Local Area Network (LAN) Topology

A Local Area Network (LAN) is a common computer and telecommunications network. This section covers the Network Layer's role in this type of network.

A *topology* is a standard method of connecting systems on a network: It is the design layout of a network. Before a network is installed, it is critical to select a topology that is appropriate to the intended use of the network. Selecting the best topology for an installation requires several questions to be answered:

- What applications will be used on the network?
- What types of hosts and file servers are to be connected?
- Will the network be connected to other networks?
- Will the network have mission critical applications?
- Is data transmission speed important?
- What network security is needed?
- What is the anticipated growth in the use of the network?

Local Area Network Characteristics

There are four characteristics of a local area network:

- Single Ownership

- — Owned by a single company or organization

- — Pertains to network control

- ▪ Small geographic regions

- — Limited to a single building or small group of buildings

- — Largest distance between two stations is 05 to 50 km

- ▪ Low error rate

- — Acceptable error rate is less than 1 error in every 108 to 1,011 bits

- — 1,000 times lower error rate than the telephone network

- ▪ High data rate

- — Bit transfer rate in excess of 1 Mbps

- — Can have a high data rate of 100 Mbps

LANs provide general interconnectivity between devices. Connections between unlike devices require interfaces that may use an inefficient software or hardware strategy. A network design that provides "bandwidth to burn" and low error rates often sacrifices the ratio of overhead bits to data bits for these interface connections.

LANs do not provide end-to-end communication; they only provide an interface to the communications media. The better topologies of device LAN interconnections do not guarantee that a system will have useful communication. Additional layers of protocol are required above the LAN interface for end-to-end communication.

Common Physical Layer devices used in telecommunication topologies are:

- ▪ Routers

- — Internetwork compatible

- — Composed of a computer and packet switch

- — Connect multiple LANs and complex networks

- — Use network layer addresses

- — Connect interface ports

- Bridges
  - Higher speed than routers
  - Only utilize lower 2 layers
  - Connect two similar LANs
  - Provide filtering
  - Implemented:
    - Locally by connecting LANs in same building
    - Remotely, by connecting LANs across town with a private line
- Hubs
  - Routing hubs
  - Combine the functions of:
    - Media Access Unit (MAU) or wiring hub
    - Bridge
    - Router
    - Network management
    - Ethernet, token ring, FDDI, ATM
- Switches
  - Connect bandwidth for voice, data, and video communication
- Gateways
  - Connect heterogeneous networks (DEC and SNA)
  - Provide upper-layer protocol conversion
  - Hardware can be a:
    - PC running gateway software
    - Processor (VAX or X.400) running a gateway application
- Modems
- Interfaces

— Provide voice-grade services

— Use Channel Service Units/Digital Service Units (CSU/DSU), which interface T1/E1 services

— Use *Terminal Adapter/Network Termination* (TA/NT1) devices, which interface Integrated Services Digital Network (ISDN) services

- Communication servers, which concentrate dial-in and dial-out user communication

Logical Topologies

A logical topology determines how nodes are to communicate across the medium. The most common logical topologies in use today are:

- Token-passing

- Broadcast

- Bus

- Token bus

- Star

- Extended star

- Hierarchical

- Mesh

- Ring

- Fiber Distributed Data Interface (FDDI Ring)

- Fiber Distributed Data Interface II (FDDI II Ring)

TOKEN-PASSING.   A token-passing network:

- Controls network access by passing a token (electronic signal) sequentially to each node

- Upon receipt of the token, that node has permission to send data on the network

■ Upon receipt of the token, if that node does not have data to send onto the network, the node passes the token to the next node, and the process repeats itself

Figure 7.13 illustrates of a token-passing network.

Figure 7.13
Token passing topology.

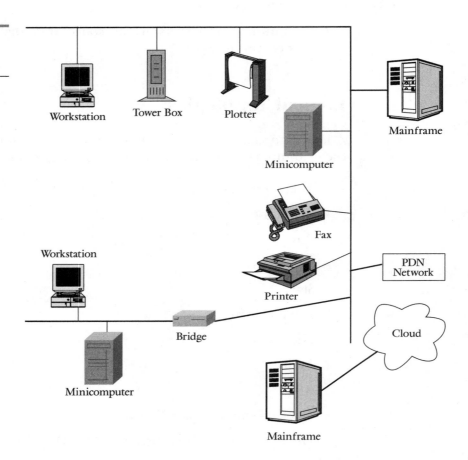

BROADCAST. In a broadcasting network:

■ Each node transmits its data to all other nodes on the network medium

■ Nodes need not be in a particular physical layout to receive the broadcast signal

- Transmission and reception is on a first-come, first-serve basis

- The Ethernet environment is used

- No routing requirement or flow control is required

Figure 7.14 illustrates a typical broadcast topology.

Figure 7.14
Broadcast topology.

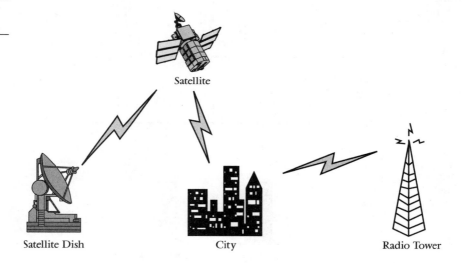

Satellite

Satellite Dish

City

Radio Tower

BUS. A bus topology uses a single backbone segment (length of cable) to which all the hosts connect directly. This bus is the main network cable or line that connects network stations. The bus topology is one of the most common "shared cable" network designs. It consists of multiple stations that are independently attached to a shared cable. The bus must end in a terminating resistance, or *terminator,* which absorbs electrical signals so that they do not bounce back and forth on the bus. Figure 7.15 is a diagram of the physical layout of a bus topology.

Bus topologies are common in the telecommunications environment because a shared-medium network fits well into the concept of a broadcast network. Stations tap onto the bus using passive devices, each of which must determine its target destination.

To ensure that only one node (workstation) transmits data at a time, a bus topology uses *collision detection;* otherwise, a *collision* will occur.

When a collision occurs, the voltage pulses from each device are both on the common bus wire at the same time; as a result, the data from both devices collide and are damaged. When collision is detected, a signal is issued to all nodes to enforce a retransmission delay in order to minimize another collision.

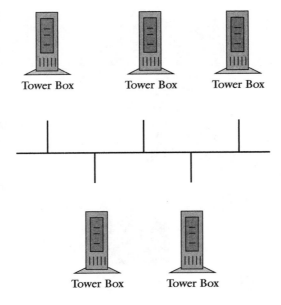

Figure 7.15
Bus topology.

Advantages of a bus topology are:

- Low implementation cost
- Well-established technology
- Simple node connections to the shared medium
- Failure of stations does not affect network
- Uses the least amount of cable of all the topologies

Disadvantage of a bus topology are:

- Unable to use fiber optic cable
- Management costs are high

- Difficult to isolate a single malfunctioning node

- One defective node can take down an entire network

TOKEN BUS. A token bus network topology uses an access scheme in which all stations attached to the bus listen for a token. Stations utilize a token when they have data that must be transmitted. The physical topology design layout of the token bus is the same as the typical network bus described earlier.

STAR TOPOLOGY. The star is the oldest form of network topology. It was introduced with the digital and analog switching devices used in telephone systems. A star topology consists of multiple peripheral nodes, having distributed control. All stations have a direct, point-to-point connection to the central node and are equally responsible for establishing a connection. The central node performs like a physical switch. See Figure 7.16.

Figure 7.16
Star topology.

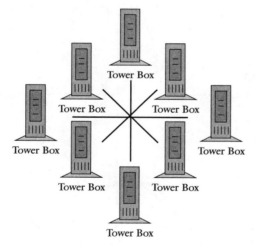

The central node or switch may be circuit oriented (i.e., PBX), packet oriented (i.e., DATAKIT), or a computerized branch exchange (i.e., CBX). The biggest advantage of a star topology is its capability to integrate voice and data. A weakness of the star topology is that it does not check for compatibility and does not provide protocol conversion.

The star network is easy to manage because malfunctioning peripheral nodes or damaged wiring can be isolated from the network without affecting service to other peripheral nodes. A disadvantage of the star is that central switch failure leads to a single point of failure.

The star's initial cost is the highest of all LAN topologies. The initial high cost is caused by the expense of the central switch and the expense of wiring a pair of cables from each peripheral node to the central switch.

EXTENDED STAR. An extended star topology links individual stars together by linking the switches, thus extending the length and size of the network. See Figure 7.17.

Figure 7.17
Extended star
topology.

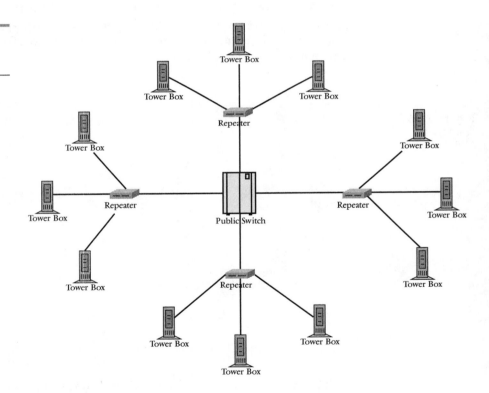

HIERARCHICAL. The hierarchical topology is similar to an extended star but instead of linking switches together, the system is linked to a computer that controls the telecommunications traffic on the topology. See Figure 7.18.

Figure 7.18
Hierarchical
topology.

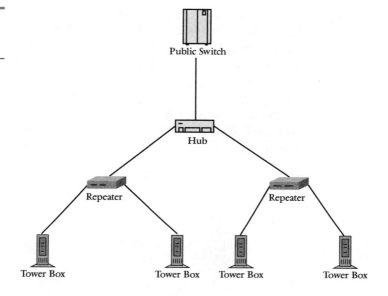

MESH. A mesh topology occurs when there are no breaks in communications; each node has its own connections to all other nodes. See Figure 7.19.

Figure 7.19
Mesh topology.

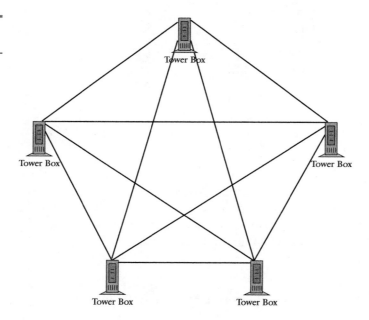

RING. The ring topology has no endpoint or terminators. All nodes are connected via a circular closed set of point-to-point links—the first node connects to the next node and the last node to the first node. The layout is a continuous loop of cable to which the network nodes are attached. The ring is classified as a network access method and topology in which a token is passed from station to station in sequential order. Each station on a ring is connected via an active tap that is typically a repeater. Bits come in on the receive side of the tap and leave on the transmitter side of the tap. See Figure 7.20.

Figure 7.20
Ring topology.

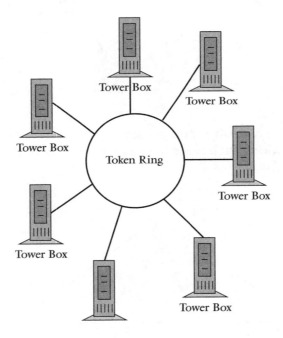

Rings with distributed control are common in the telecommunications environment. All nodes on the ring participate equally in the access control procedures, using a scheme to ensure that two stations do not transmit on the ring at the same time. The advantages and disadvantages of a ring network are:

- Advantages:
    - Fiber optic cable may be used

— May be used across large geographic distances

— Transmits signals over long distances

— Handles high-volume network traffic

— Easy to locate a defective node

- Disadvantage:

— Relatively expensive to implement

— Reliability is a potential problem

— Uses a lot of cable

Fiber Distributed Data Interface (FDDI Ring)

FDDI is a high-speed (100Mb/s) ring technology that consists of two loops for redundant data transmission in opposite directions. The underlying medium is fiber optic cable, and the topology uses dual-attached, counter-rotating token rings. See Figure 7.21.

Figure 7.21
FDDI ring topology.

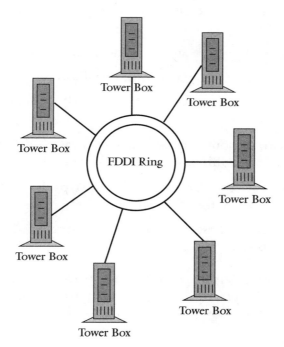

Fiber Distributed Data
Interface II (FDDI II Ring)

FDDI II Ring is a newer standard that carries multimedia signals at 100 Mbps. Otherwise, the architectural design is the same as plain FDDI.

Contention Access Method (CAM)

Originally, the *Contention Access Method* (CAM) system was called *Pure Aloha*. It was named for its first implementation in the ALOHANET, built at the University of Hawaii in the late 1960s. In today's environment, Aloha or Contention Access Method (CAM) is used on bus LANs.

The Pure Aloha contention scheme allows all stations on the network to transmit when needed, so that collision of packets is a risk, and packets are destroyed during collision.

Slotted Aloha is an enhancement of Pure Aloha:

- If transmitter sends a packet only at clock signal, the packets will be safe

    — Clock signal indicates the beginning of a packet interval

    — Packets do not encounter collisions

    — Station waits to transmit until the next packet boundary transmits

- If two stations become ready to transmit during the same packet interval a collision may occur without the clock signal

    — Packets transmit at the next packet interval and collide, destroying only a single packet

    — If the first bit is transmitted successfully, the entire packet is safe

Contention buses operate at high speeds to shorten the packet time; therefore the probability of two stations becoming ready within the same slot is very low.

To enhance the Aloha contention schemes, all stations utilize a *Listen Before Talking* (LBT) scheme called *Carrier Sense Multiple Access* (CSMA). CSMA access schemes reduce the number of collisions by not transmitting while the line is in use. If a station detects an idle channel, it transmits. If the channel is busy, the station waits and listens again later on. In this way, the need for fixed packet sizes is eliminated.

To further reduce collision, a scheme called *CSMA with Collision Detection* (CSMA/CD) can be used. In CSMA/CD systems, a station continues to sense the channel during transmission using a *Listen While Talking* (LWT) scheme. If a collision is detected, the transmission is suspended. CSMA/CD is the access mechanism used in Ethernet networks.

Contention systems are hindered by an increased number of stations on the cable and increased amounts of data traffic. Each station on the bus handles one outstanding packet at a time, so the critical issue on the bus is the number of contending stations, not the number of packets that the station wants to send. It is theoretically possible that some stations will be blocked forever because of an unbounded maximum delay.

Network Layer (Level 3) Protocols

A protocol is defined as a set of agreed upon rules that facilitate communication. So far we have discussed two levels of protocols:

- Physical Layer (Level 1) protocols
    - Describe how a DTE plugs into a data circuit terminator
- Data Link Layer (Level 2) protocols
    - Structure of the Data Link Protocol and assures that the physical communication path will appear to be error-free to higher layers.

This chapter discusses the Network Layer (Level 3) protocols that are required to obtain services from a computer network. Figure 8.1 is a OSI—Telecommunications comparison chart that reviews how the Network Layer (Level 3) relates to the other levels for communication.

Figure 8.1
Network layer of the OSI Telecommunications Reference model.

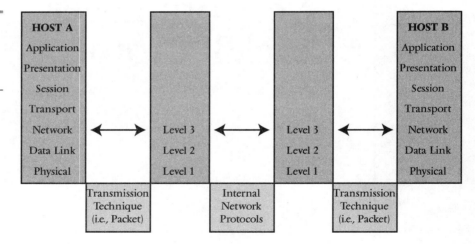

Voice Network Protocols

To utilize a voice telephone network, certain procedures defined by the network must be followed. For example, let's go over the procedure of setting signals to establish a telephone call:

- Removal of the telephone receiver from the cradle sends one signal to the telephone network as a request for service

- The network searches for facilities to begin setting up the call

- When sufficient facilities have been determined, a specific signal is sent to the user (dial tone) to inform the user that she may begin dialing

- While the user is dialing the telephone number, she utilizes specific frequencies in signaling the network

- Based on these frequencies, the network agrees to send call-progress signals to inform the user of the status of her call:

  — Busy tone

  — Reorder tone

  — Ringing tone

Four network protocols have been extracted from the above procedures:

- Taking the switch hook off of the cradle

- Receipt of dial tone

- Multifrequencies received when dialing the number

- Call progress signals

Communication Architectures

Communication network architecture describes the specific set of functions and protocols needed to accomplish end-user-to-end-user interoperability. Communication network architecture provides the specific layered protocols and interfaces required to provide connectivity and interoperability between end users and applications.

Network architecture describes the elements and protocols (OSI layers 1—3) that have a responsibility for establishing these end-to-end connections (see Table 8.1). A customer's total communication architecture may include a subnetwork with differing subnetwork architectures (i.e., TCP/IP, X.25, ATM, etc.) which may be implemented by the customer or a telecommunications network provider.

A telecommunications network provider constantly works to develop and deliver data network products and services that add value for clients.

TABLE 8.1

Voice and Data Network Layer Comparisons

	Voice	Data
Layer 3	Called address	Called address
	Billing number	Packet size
	Class of service	Datagram/virtual circuit
		Internet Protocol (IP)
		X.25 Packet Level Protocol (PLP)
Layer 2	Ground start	Link address
	Loop start	HDX/FDX protocol
	E&M	Link window size
	etc.	Link timer values
Layer 1	−48V	EIA-232dc
	300—3300 Hz Analog	CCITT V.35
	T1-CXR	X.21
	etc.	HDX/FDX facilities
		etc.

Network Architectures

Prior to 1982, network services primarily interfaced with the customer at the Physical Layer (Level 1). With additional services—X.25, Frame Relay—deployment and interfaces became an issue. Network performance can be a sensitive situation depending on user's tolerance and the cost of delayed access or blocked calls to network services.

Network architecture designs must be based on the differing network performance characteristics and application performance requirements. An analysis of the applications that will be using the network is required. Questions providing insight into an application's needs must address:

- Response time requirements

- Utilization

- Throughput

- Security

- Reliability

To determine the specific architecture protocols to be supported, the following questions must be addressed:

- Does the interfacing end-user devices or systems utilize the same protocols?

- Is the data link protocol asynchronous or synchronous?

- Is protocol conversion a possibility?

- Is network recovery necessary?

These questions must be answered, quantified, and factored in when developing and cost-justifying optimal network solutions. Table 8.2 illustrates a comparison of common network architectures in use today.

TABLE 8.2

Required Network
Architecture
Components

Hardware	Layers	Description
Gateways	Possibly all 7 layers	Perform protocol conversion between dissimilar networks
Routers	Layer 3	Devices which make routing decisions and may implement Layer 1—3 gateway functions (protocol conversions)
Bridges	Layer 2	Devices which normally connect LANs and may perform link layer address filtering and/or translations

When acquiring an understanding of a customer's network architecture, an architecture comparison is helpful to a common reference. The Open Systems Interconnect (OSI) is a good common architecture reference for comparison purposes.

Homogeneous architectures, such as SNA or DNA, define all the services required and provided under that specific architecture for end-to-end communications to occur. Few customers have a homogeneous architecture. A homogeneous architecture is only seen when a single vendor's equipment and software are deployed.

Most customers operate in a multivendor environment, so that it is necessary to provide interfaces from one environment to the other. To provide communication, the following solutions must be reviewed:

- A *gateway* service must be provided to resolve protocol incompatibilities.

- It may become necessary to use an intervening network that is not a defined subset of a homogeneous communications architecture for interconnecting similar to dissimilar networks or network components.

- Protocols implemented by the intervening network must be addressed and their impact upon the connected systems must be understood.

■ Several chained layer conversions may be necessary, whereas only one conversion is necessary for end-to-end layers.

Figure 8.2 illustrates these required network architecture components.

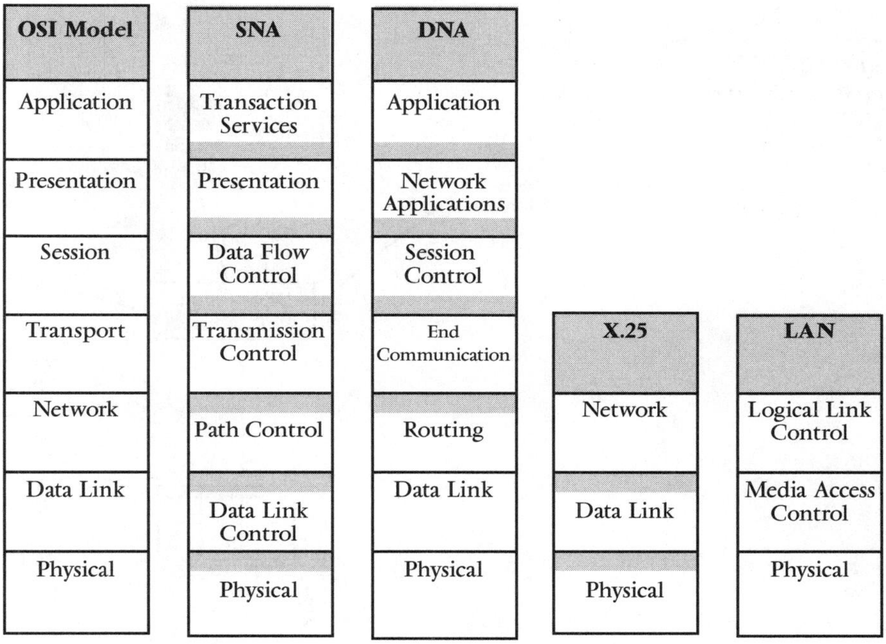

Figure 8.2
Common network architectures and protocols.

OSI Model	SNA	DNA	X.25	LAN
Application	Transaction Services	Application		
Presentation	Presentation	Network Applications		
Session	Data Flow Control	Session Control		
Transport	Transmission Control	End Communication	X.25	LAN
Network	Path Control	Routing	Network	Logical Link Control
Data Link	Data Link Control	Data Link	Data Link	Media Access Control
Physical	Physical	Physical	Physical	Physical

BiSync Protocol

BiSync is an outdated protocol, but there are still customer network architectures based on this 3270 bisynchronous protocol. Prior to SNA, the BiSync protocol was based on networking the workplace environment where workers were more or less specialized in their work functions (i.e., one worker handled accounts receivable; one worker handled accounts payable, etc.). See Figure 8.3.

Each worker was given a data terminal, and physical definitions provided access to a single application. Because jobs were so specialized, it was assumed that the work location (desk) area was also specialized. It simply was not envisioned that a single person would need to have access to multiple applications from a single location (terminal, desk, etc.). Neither communications software nor hardware had progressed to the point that switching between applications was feasible. If a single worker would perform two duties, the worker was given an additional terminal attached to that additional application.

Figure 8.3
The BiSync protocol.

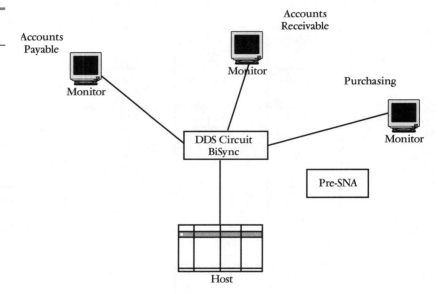

The Basic Telecommunications Access Method (BTAM) (Figure 8.4) along with BiSync, is a data link protocol that provided early users of mainframe computers with relatively little network flexibility. The BTAM environment provided opportunities for:

- The sale of analog and digital private line services
- Packet switching
    - Ability to switch between applications from a single terminal
    - Reducing and delaying the need to make costly SNA conversions
- SecureNET or NRS risk management
- BDS opportunities for placing on premises coaxial cable

Figure 8.4
A BiSync and BTAM architecture.

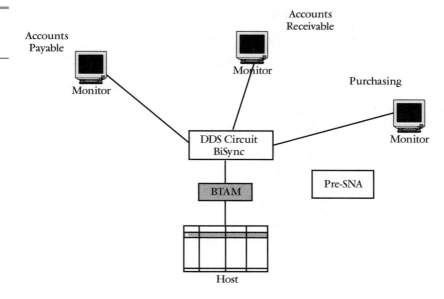

Systems Network Architecture (SNA)

Systems Network Architecture (SNA) (Figure 8.5) evolved within a business environment. SNA's logical approach to computing was to provide a single, powerful, centrally programmed computer that served repetitive business applications and large numbers of end users with common intertwining needs.

Systems Network Architecture (SNA) is a complete communications architecture, as is its predecessor, BTAM/BiSync, but is significantly more sophisticated. SNA has the ability to perform the following functions:

- Specify the complete set of protocols that must be used to communicate in this environment

- Use rules to define a limited set of end users that the network is willing to link

- Be aware of these users based on logical unit or logical entity definitions

- Refer to logical units and logical entities collectively as *Network Addressable Units* (NAUs)

- Define three types of network addressable units:

    — Logical Units

    — Physical Units

    — System Services Control Point (SSCP)

- Describe relationships between the various Network Addressable Units (NAUs)

- Define the protocols governing the exchange of information between them

Systems Network Architecture (SNA) (Figure 8.5) and *Virtual Telecommunications Access Method* (VTAM) provided an improvement over BTAM/BiSync by allowing end users to:

Figure 8.5
Systems Network
Architecture (SNA).

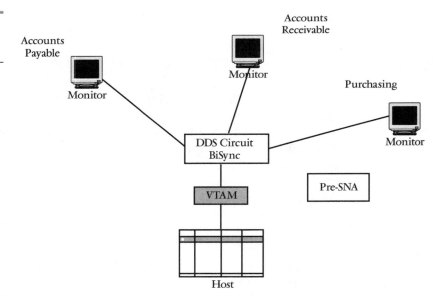

- Request connections to multiple applications from a single terminal

- Relieve application programmers of having to write code within the applications to control sessions

- Provide a centralized flexible means to establish, maintain, and terminate sessions

- Contain a hierarchical architecture in which control of all resources, knowledge of all routes, and establishment of all connections resides within a centralized host processor (the VTAM)

Digital Network Architecture (DNA)

Digital Network Architecture (DNA) grew in an environment of technical users. This environment dictated that computing power was close to the researcher, who tended to develop and work with specialized programs. To manage such an environment, smaller groups of users required direct control of their computing resources. Minicomputers became the logical choices.

In the late 1970s, an approach was established to use minicomputers for parallel applications processing and file transfer over peer-to-peer dynamic network architectures. In today's environment, a distributed peer-to-peer architecture is used in office automation environments. The connection of a common user interface (logical) to an easily implemented network, which supports a wide range of topologies and communications facilities, is the ultimate goal for ease of maintenance and usability.

The Digital Network Architecture (DNA) has evolved in phases:

- Phase 1

    — Implemented in the late 1970s

    — Used to interconnect minicomputers

    — Required point-to-point connections between minicomputers

    — File transfers between noninterconnected systems required sequential sessions

    — The following transmissions were available:

        - Synchronous and asynchronous

        - Serial and parallel physical interfaces

- Phase II

  — Implemented in late 1970s, after Phase I

  — Interconnection of minicomputers became transparent to network users

- Phase III

  — Implemented in early 1980s

  — Added routing capabilities

  — Computer served as routers in the network

  — Routing table updates were based on network metrics

  — Eliminated of the need to establish sequential sessions for file transfers among nonadjacent computers

- Phase IV

  — Implemented in mid-1980s

  — Added of the following protocols:

    - Ethernet

    - LAPB

    - SNA

    - CCITT X.25 gateways

  — Used VT100 terminals (also known as *dumb terminals*)

  — Allowed transparent connections from users to remote nodes

- Current implementation

  — Supports CCITT-defined OSI protocols while retaining current compatibility

  — Encourages third-party development of applications software

X.25 Protocol Networks

The International Telegraph and Telephone Consultative Committee (CCITT) is one of the committees of the International Telecommunication Union (ITU), an agency of the United Nations.

CCITT recommendation X.25 is a description of the network protocol to be used by public data networks (PDNs). It describes the syntax (form) and semantics (meaning) of the packets that are exchanged with the network. X.25 is a careful specification of the exact structure of these packets. Recommendation X.25 was developed and introduced in 1976 as user—Packet Switched Public Data Network (PSPDN) interface. Recommendation X.75 was introduced in 1980 as PSPDN—PSPDN interface. The CCITT X-series recommendations address the access to digital PDNs.

X.25 does not support the same speed as newer technologies, but it offers reliable data communication over networks in countries lacking in technology development. In the United States, packet switching is mainly used for internal telephone service.

CCITT recommendations are published every four years, and their book covers are color-coded. For example, 1980 recommendations have yellow book covers and 1984 recommendations have red book covers. It is common to refer to these specifications by color—Red Book specifications, Green Book, and the like.

CCITT Recommendation X.25 Interface Overview

X.25 defines the interface between Data Terminal Equipment (DTE) and Data Circuit Equipment (DCE). The DTE represents the end user, or host system. The DCE represents the boundary node of the Packet Switched Public Data Network (PSPDN)—in other works, the DCE is the point of access into the network (see Figure 9.1).

Recommendation X.25 does not address the subnetwork connection between the DCEs. The network's internal architecture is up to the PSPDN implementor and is totally transparent to the DTEs.

Figure 9.1
CCITT
recommendation
X.25 interface.

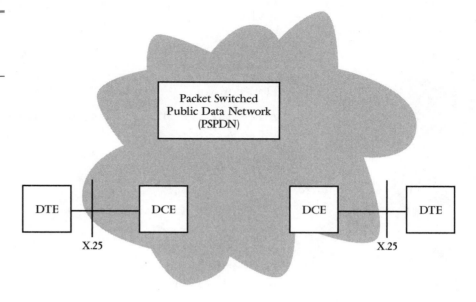

Figure 9.1
CCITT
recommendation
X.25 interface.

X.25 Reference Model

CCITT Recommendations define three X.25 levels for communications between the Data Terminal Equipment (DTE) and the Data Circuit Equipment (DCE). Each level addresses a different OSI layer:

- Physical Layer (Level 1)

  — Physical connections

  — Addresses bits

  — Timing

  — Control signals

  — Interface X.21, X.21 BIS used

  — Similar to the RS-232C standard for serial communications

- Data Link Layer (Level 2)

  — Point-to-point connection between the DTE and the network

    — Address frame

    — Data transfer

    — Error checking

    — Flow control

    — Access level is Link Access Protocol-Balanced (LAP-B)

    — Address packets

    — X.25 packet switching format

- Network Layer (Level 3)

    — Address packets

    — X.25 packet switching format

    — Defines how a packet terminal interfaces with a packet-switched network

Figure 9.2 illustrates how the X.25 is *commonly* thought of relating to the OSI Reference Model.

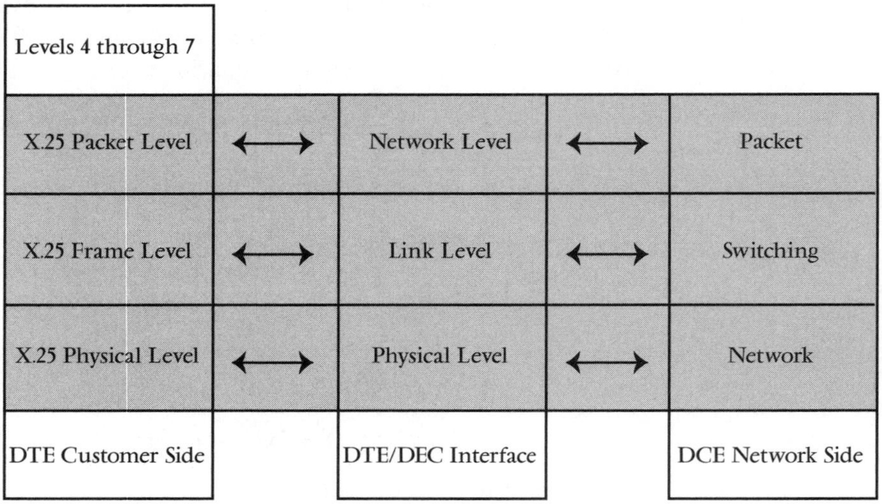

Figure 9.2
OSI model and recommendation X.25 comparison.

Levels 4 through 7				
X.25 Packet Level	⟷	Network Level	⟷	Packet
X.25 Frame Level	⟷	Link Level	⟷	Switching
X.25 Physical Level	⟷	Physical Level	⟷	Network
DTE Customer Side		DTE/DEC Interface		DCE Network Side

I use the word *commonly,* because the relationship is commonly thought of as being straight across.

In reality the OSI model is designed to deal generically with the most general types of networks, whereas X.25 deals with a very specific application. Table 9.1 shows the specific applications addressed by the X.25 protocol.

TABLE 9.1

Specific
Applications
of X.25

OSI X.25 Application Layers	Descriptions
X.25 Physical Level Subset of OSI Physical Layer	Dedicated Point-to-point Serial Synchronous Full-duplex channel
OSI Physical Layer	Parallel vs. serial Half-duplex vs. full-duplex vs. simplex Switched access vs. leased access Point-to-point vs. multipoint Synchronous vs. asynchronous
X.25 Link Level Subset of OSI Data Link Layer	Ensures error-free communication over the link Less general than OSI Data Link Layer because data link is narrowly defined by the X.25 Physical Level
OSI Data Link Layer	Ensures error-free communication over the link. Broader than X.25 Link Level because data link is specifically defined to certain protocols by the OSI Physical Layer.
X.25 Packet Level Protocol (PLP) Subset of OSI Network Layer	Routing is not necessary over a point-to-point channel. Flow control

continued on next page

TABLE 9.1

Specific
Applications
of X.25
(Continued)

OSI X.25 Application Layers	Descriptions
OSI Network Layer	Routing is necessary when it is not a point-to-point channel
	Flow control
X.25 Packet Level Protocol (PLP) Subset of OSI Network Layer	End-to-end functions, such as segmentation of messages into packets and reassembly of packets into messages
	End-to-end data packet acknowledgment
OSI Transport Layer	End-to-end functions, such as segmentation of messages into packets and reassembly of packets into messages
	End-to-end data packet acknowledgment

X.25 refers to only the lower three layers of the OSI model; levels 4 through 7, the end-to-end layers, are beyond the scope of the X.25 user—network interface (Figure 9.3).

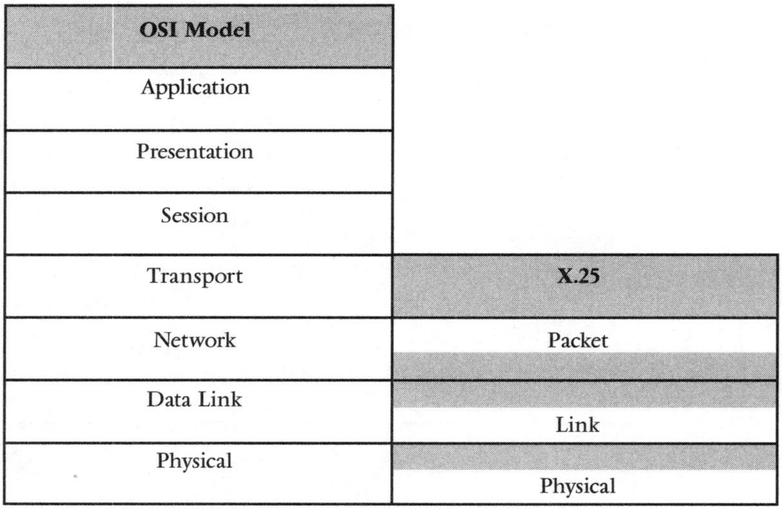

Figure 9.3
X.25 refers to the
lower four layers of
the OSI Reference
Model.

OSI Model	X.25
Application	
Presentation	
Session	
Transport	X.25
Network	Packet
Data Link	Link
Physical	Physical

X.25 Physical Layer

The Physical Layer of the OSI Reference Model defines the physical interface between two adjacent devices. The physical link between a DTE and DCE is a dedicated, synchronous, serial, point-to-point full-duplex channel. The X.25 Physical Level provides a subset of the functions defined by the Physical Layer of the OSI model. X.25 also requires a dedicated, synchronous, serial, point-to-point, full-duplex circuit between the user (DTE) and network (DCE).

Recommendation X.25 specifies that the Physical Layer conform to CCITT Recommendation X.21 or X.21 bis. Other interfaces may also be used, including V.24, V35, RS-232-C, RS-449, RS-422-A (V.11, X.27), and RS-423-A (V.10, X.26).

X.25 Link Layer

The OSI Data Link Layer provides error-free communication between two adjacent devices. Although the media connecting two devices may not be perfect, the Data Link Layer is responsible for finding and correcting all bit errors on the line.

The X.25 Link Layer provides the same functionality of the OSI Data Link Layer. The X.25 Link Layer is not a general protocol. For example, no mechanism exists in X.25 to turn the line around for half-duplex communication because the physical link must be full-duplex. Nor are there the polling and selecting facilities required for multidrop configurations.

X.25 has two Link Layer procedures:

- Link Access Procedure (LAP)

  — Part of the original X.25 recommendation in 1976

  — New implementation discouraged

- Link Access Procedure Balanced (LAPB)

  — Introduced in 1980

  — Since LAPB was introduced, the CCITT has discouraged new LAP implementations in favor of LAPB

  — A bit-oriented protocol

— Uses frame format, error-detection algorithm, and sequential frame delivery that is similar to the following protocols:

- Synchronous Data Link Control (SDLC)

- ANSI Advanced Data Communications Control Procedure (ADCCP)

- ISO High-level Data Link Control (HDLC)

LINK ACCESS PROTOCOL-BALANCED (LAP-B). The Link Access Protocol-Balanced (LAP-B) is responsible for:

- Initiating communication between DTEs and DCEs

- Ensuring that frames arrive at the receiving node in the correct sequence

- Verifying that receiving packets are error free

- Providing one of three primary unnumbered command collisions:

  — Unnumbered command collision

    - When the same commands are used, both stations respond with Unnumbered Acknowledgment (UA)

    - When the commands are not the same, a Disconnected Mode (DM) response occurs

  — Set Asynchronized Balanced Mode (SABM)

    - Either end can initialize a SABM

    - When received, UA should be sent as a response

  — Disconnect Mode (DM)

    - If the DCE or DTE cannot initialize the connection and a SABM command has been initiated a Disconnect Mode (DM) occurs. (An UA command would normally be sent to acknowledge the SABM command.)

- Network Layer (Level 3) or Packet Level Protocol (PLP)

LAP-B FRAME STRUCTURE. The LAP-B frame structure is illustrated in Table 9.2.

TABLE 9.2

OAP-B Frame
Structure

Frame Fields	Description
FLAG	Bit pattern 01111110 is used to delimit the beginning and end of the frame (8 bits)
	Zero-bit insertion (bit stuffing) is used to ensure that a FLAG bit pattern does not appear between real FLAGs
	Bit stuffing forces the transmitter to either send a 0 after every group of five contiguous 1 bits or send a FLAG
ADDRESS	Used in X.25 to differentiate between commands and responses (8 bits)
CONTROL	Indicates type of the frame (Information, Supervisory, or Unnumbered)
	Information frames carry data and sequence number (8 or 16 bits)
INFO	Contains the Link Layer data Field can carry any number of bits
	Information field of an Information frame contains a single packet from the X.25 Packet Layer
FCS	Frame Check Sequence fields contain the remainder from a Cyclic Redundancy Check (CRC) calculation to ensure that the frame has no bit errors
	The CRC-CCITT polynomial is used (16 bits)

CONTROL FIELD IN LAPB FRAME. LAPB defines three types of frames in the control field:

- Information (I-frame)
  - Exchanges data
  - X.25 packets comprise the Information field
- Supervisory (S-frame)

- — Controls the exchange of Information frames

- — Acknowledges Information frames

- — Stops the transmitter from sending more frames

- — Requests retransmission of Information frames

- Unnumbered (U-frame)

  - — Controls the operation of the link

X.25 Packet Link Protocol

The X.25 Packet Link Protocol (PLP) provides a subset of the functions defined by the OSI Network Layer. One of the Network Layer's functions is routing. X.25 provides a point-to-point interface so that routing is not needed. Both the Network Layer and X.25 provide flow control. The Packet Protocol:

- Provides message fragmentation for transmission; reassembles the message at the arriving end

- Provides a concentration facility

- Deals with a single physical and logical connection in the Physical and Link Levels of X.25

- Provides up to 4,095 logical channels on a single physical channel

- Supports a virtual circuit

- Does not support datagram service

- Provides rules for the establishment of virtual calls

- At virtual call setup, network assigns a logical channel number to the call and all packets refer only to the logical channel

There are three modes of the X.25 Packet Link Protocol:

- Switched Virtual Circuits (SVC)

  - — Analogous to a dial-up telephone call

    - - Call request is made

- Call is put through if network resources are available

- Upon completion of call, resources allocated for call released

- If resources not available, call is not placed

— Two-way transmission path is established from node to node

— Logical connection is established only for the duration of the data transmission

— Upon data transmission completion, the logical connection becomes available to other nodes

- Permanent Virtual Circuits (PVC)

— Similar to a leased line

— Logical connection remains connected for instant transmission at all times

— Resources are reserved and always available

— Connection remains in place even when data transmission stops

— Establishment of PVCs is performed by the network at subscription time

- Datagrams

— Packaged data sent without establishing a communication channel

— Packets may take different routes so they may reach their destination at different times

— Not used on international networks, but are included in the CCITT specifications for the Internet

Packet-Switching
Network Protocols

A packet-switching network utilizes user-to-network and user-to-user protocols. The network specifies certain requirements concerning the means by which it addresses other DTEs. Packet-switching networks are subject to congestion when traffic becomes heavy enough that long queues develop. Packet networks use some form of congestion control and send signals to tell users when they must reduce their traffic load on the network. Message formats and user responses are all part of the network protocol. When a packet is delivered at the remote DTE, the network has fulfilled its responsibility; however, additional addressing information is required for the remote DTE to determine to which program the data is destined.

Basic required user-to-network protocols:

- Address the DTEs

- Provide congestion control of traffic

- Detect and correct errors

- Provide a physical connection

With today's need for security requirements, if a transmitter encrypts data before sending it through the network, the receiver must know how to decode the arriving message. This requires an agreement between the two communicating DTEs.

User-to-network protocols for encryption:

- Identify of target program

- Control flow of data

- Provide encryption

Packet Switching and X.25

Packet switching is the X.25 Network Layer (Layer 3) of the OSI Reference Model. Packet switching most commonly follows the X.25 standards for bit-oriented protocols. A packet switching network consist of a number of network nodes linked together with point-to-point data links. These nodes perform basically three functions:

- Establish a connection between telecommunication network equipment and the equipment using the network

- Direct switching operations by determining the route

- Transmit data from one network to another network

The X.25 standard defines communication between Data Terminal Equipment (DTE) and Data Communications Equipment (DCE). In packet switching, the DTE can be a computer or a host machine and DCE can be the packet-switching node. The DTE is attached to a *Packet Assembler and Disassembler* (PAD). The PAD offers the following functionality:

- Translates data from DTE format into X.25 format

- Translates X.25 format into DTE format

- Provides extensive error detection and correction

- Can send out data from several DTEs at the same time

Data Transmission

Packet switching is a store-and-forward switching method—messages are stored and then forwarded to their destination. For transmission efficiency, packet switching messages are divided into smaller packets.

A simple way to think of packets is to compare them to a mailing envelope with a destination address, source address, and packet number written on the face. This addressing does not interfere with the data inside the envelope. In fact, the data could be missing or in the wrong envelope, and the packet will still be delivered according to envelope instructions.

A packet switching network (Figure 10.1) divides the data into blocks that are called packets. Each packet is addressed with the destination address, source address, and other control information. At each node that the packet reaches along the way to the source destination, the address is analyzed, just as in a postal system when an envelope tries to reach its addressed destination.

Using the packet connection procedure:

- A DTE divides user data into packets consisting of control information (destination address, start and stop bits, etc.)

- Packets are forwarded on to the DCE

- If multiple DCEs are required for a packet to reach its final destination, each DCE examines the destination address and selects the next network node on the route

- When the node makes its selection for the best route, chances are it will not always choose the same route, due to some routes becoming congested

- Because of the various routes used, packets can arrive at their destination out of sequence

- A Packet Assembly and Disassembly (PAD) is used to reassemble packets into their original structure

- The PAD can be located on the user's or the packet switching network

Figure 10.1
A packet-switching network.

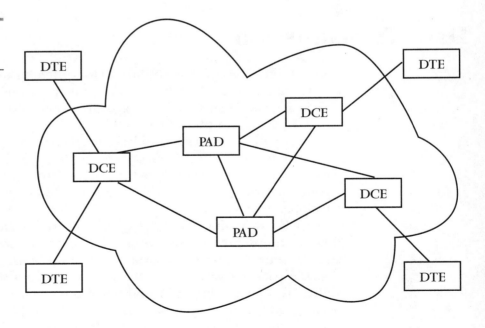

To establish a connection between a user and a packet switching network:

- Wiring or cable may be used if the network computer is located on site

- Modem and wideband line connection may be used

- Dial-up voice-grade connection may be used

Packet Switching Functions

The main functions of packet switching are to:

- Transfer data by creating control and data packets

- Connect and supervise circuits to remote DTEs

- Implement LAP-B procedures to transfer data across the DTE—DCE interface

- Ensure no loss of data through flow control

- Ensures correct addressing of data and control packets

Control Functions

The major control functions provided by the packet switching network are:

- Call Request
  - Transmission of a Call Request packet to establish a virtual call

- Incoming Call
  - Conversion of the Call Request packet into an Incoming Call packet, which is sent to its destination

- Call Accepted

- — A positive response to an Incoming Call packet

- Call Connected

  - — Conversion of the Call Accepted packet into a Call Connected packet

- Clear Indication

  - — Destination station is unable to accept an Incoming Call packet

  - — The Incoming Call packet gives a reason for refusing the call

- Clear Request

  - — Station wants to disconnect virtual call

- Clear Confirmation

  - — A Clear Request receives a positive acknowledgment

  - — Transmission is sent as a final step in disconnecting a virtual call

Packet Switching Virtual Circuit Service

A packet switching virtual circuit is a network facility that appears to be a full end-to-end connection circuit, in contrast to a physical circuit. A virtual circuit is memory-mapped and can be shared by many users.

Two types of virtual circuits are provided by the packet switched virtual service:

- Permanent Virtual Circuit (PVC)

  - — Provides a permanent physical connection to network computers

  - — Defined by the subscriber to the Postal, Telegraph, and Telephone (PTT) authorities maintaining the network at subscription time

- Switched Virtual Circuit (SVC)

    — Establishes connection for exchanging messages

    — Sets up dynamically

    — Breaks connection upon completion of message

When using either a Permanent Virtual Circuit (PVC) or a Switched Virtual Circuit (SVC), the transmission connection is transparent to the user.

Virtual Circuits Networks and Datagrams

A packet-switching network has two basic approaches for data transmission:

- Virtual circuits networks
- Datagrams

Virtual circuit networks are also called *connection-based networks.* A user asks the network to establish a path to a destination and to associate a short identifier with that path. Subsequent packets need only reference the path number. Delivery is in sequence and is guaranteed.

A mandatory set of rules must be followed by a virtual circuit to transmit data:

- Virtual circuit network requires establishment of a connection between the user and the remote DTE prior to sending data

- User provides a short connection identifier prior to establishing the connection

- Upon indication of connection establishment, data packets are transmitted with only the connection identifier and not the full destination address

- Identifiers are shorter in length than full addresses, which saves on overhead

- Virtual circuit networks also provide guarantees concerning the sequential delivery of transmitted packets

A virtual circuit network is not a circuit network. A connection in a virtual circuit network generally consists of a collection of entries in node memories that associate the short identifier on a packet with the path it is to follow. All links are still shared; there are no dedicated facilities. The term virtual circuit is used because the connection procedure is reminiscent of the connection required in a circuit network.

A *datagram network* is designed to provide maximum throughput to its users. In a datagram network:

- Every packet must carry a full source and destination address

- No setup is required for a connection

- No guarantees are provided concerning the delivery of packets

- No guarantee of the packet's delivery in sequential order is provided

- Drastic action is taken to assure that delays are minimized; may include discarding packets when traffic becomes heavy

- Responsibility is placed on the higher layers for correcting any errors that occur as a consequence of the network's action

- Most application programs are not suited to function in a datagram environment because most application programs expect data that they send (or request) will be guaranteed to be delivered

- Some end-user software can provide the appearance of virtual circuits to application programs even when the underlying network layer is of the datagram variety

Table 10.1 compares virtual circuit networks and datagram networks.

TABLE 10.1

Comparison
between a Virtual
Circuit Network
and a Datagram
Network

Virtual Circuit Network	Datagram Network
Connection based network	Nonconnection based network
Network established a short identifier destination path. Subsequent packets need only reference the path number	Every packet must carry a full source and destination address
Requires setup time	No setup time
Guarantees delivery of packets	No guarantee for delivery of packets
Guarantees sequential order or packets	No guarantee for sequential order of packets
Ideal for most applications	Works well with few applications

So that applications can obtain virtual circuit service from a datagram network, the following procedures are required:

- An end-to-end protocol is required at the receiving system to detect out-of-sequence deliveries or the absence of a packet

- Transmitted packets must be numbered in sequence

- The receiving DTE provides a set of acknowledgments and negative acknowledgments reminiscent of those provided by the Data Link Layer

- The Data Link Layer is required to correct any errors that may occur in the physical transmission

Figure 10.2 illustrates how virtual circuit service is obtained in a datagram network.

Figure 10.2
Obtaining virtual
circuit service from a
datagram network.

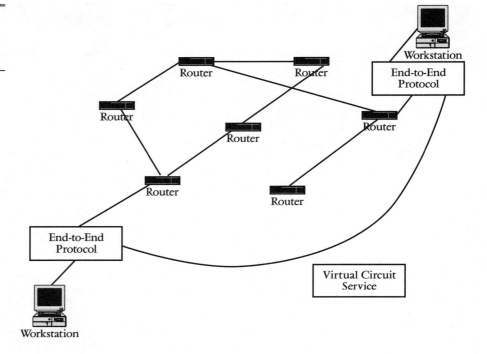

NETWORK OVERHEAD COMPARISONS. In a message network only a single address is required (Figure 10.3). This single address serves the entire data message.

Figure 10.3
Message network
overhead.

Message Network		
Address	Data	Data

In a datagram network, each packet bears the overhead once with the entire message (Figure 10.4).

Figure 10.4
Datagram network
overhead.

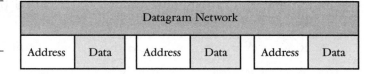

In a virtual circuit network, several overhead packets are transmitted with the data (Figure 10.5). For example, the setup message and clear message carry no user data and are overhead entities. However, to reduce overhead cost, the individual data packets carry a short identifier.

Figure 10.5
Virtual circuit
network overhead.

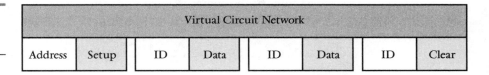

ADVANTAGES AND DISADVANTAGES OF VIRTUAL CIRCUIT OVER DATAGRAMS. One advantage of using a virtual circuit network over a datagram network is that there is a reduction of the transmission overhead associated with multiple related packets:

- Setup packet is transmitted to establish connection
- Setup packet is acknowledged so that the transmitter understands that data can be sent
- Two data packets are sent:
  — One from A to B
  — One from B to A
- Connection is broken
- During an open virtual circuit, the nodes maintain memory tables that identify the routes through the virtual circuit

There are a number of applications where datagram service is much better suited than a virtual circuit service. For instance, transaction processing is a good example of where a virtual circuit service has poor characteristics:

- A single packet is sent from A to B, followed by a single packet being sent from B to A

- Resources are used as long as the virtual circuit is open

- While the virtual circuit is open, users must pay for connection time

- To save costs, users will "take down" virtual circuits on completion of transmission

- The process of taking down the setup packet and its acknowledgment packets bring the total of packets transmitted to six

- Transmission of six packets to exchange two data packets constitutes a high overhead

Establishing Virtual Circuit Connection

Figure 10.6 details a procedure that a network might use to establish a virtual circuit connection. This procedure includes creating the table structure that is required on each router. These tables provide a tool for associating identifiers with routes on an end-to-end basis. The workstation labeled A indicates that it wishes a connection to the workstation labeled B and that it will refer to this connection as connection 4:

- Workstation A transmits a message to workstation B

- Router 1 performs a routing calculation

- Router 1 determines the best current path to workstation B is via router 2

- Router 1 may change the identifier because it serves a number of workstations

- Router 1 forwards the setup packet

- One of the set up packets may already be using identifier 4 on the path from router 1 to router 2

- As long as router 1 internally equates input identifier 4 with the output identifier 8, no problem arises

- Router 2 has now received a setup request from router 1 using identifier 8 with destination address workstation B

- Router 2 performs a routing calculation

- Router 2 determines that the best route to workstation B is via router 4

- Router 2 forwards the setup packet using identifier 6 and the destination workstation B

- Router 4 receives the setup packet

- Router 4 forwards the packet to workstation B with the identifier 5

Figure 10.6
Establishing a virtual
circuit connection.

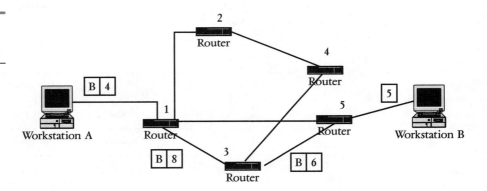

Virtual Circuit Routing Tables

Figure 10.7 shows the routing tables required in the example shown in Figure 10.6:

- Router 1 tracks the fact that it has a circuit that originates at workstation A

- Router 1 also tracks what workstation A calls identifier 4

- Router 1's table indicates that the circuit proceeds to router 2

- Router 1 changes identifier to 8 and transmits to router 2

- Router 2's table indicates that the virtual circuit originates at router 1 with an identifier of 8

- Packets proceeds to router 5 with an identifier of 6

- Router 5 indicates that the virtual circuit originates at router 2 and is called 6

- Router 5 further indicates that the circuit proceeds to workstation B with an identifier of 5

A packet originating at workstation A and carrying the identifier 4 will wind up at workstation B carrying the identifier 5. A packet originating at workstation B and carrying the identifier 5 will end up at workstation A carrying the identifier 4. The virtual circuit tables illustrated here provide for two-way communication.

Figure 10.7
Routing tables used in sample connection exercise.

Virtual Circuit Routing Tables				
Router #	**In From**	**On**	**Out To**	**On**
Router 1	Workstation A	Identifier 4	Router 2	Identifier 8
Router 2	Router 1	Identifier 8	Router 5	Identifier 6
Router 5	Router 2	Identifier 6	Workstation B	Identifier 5

Packet Structure Used in Virtual Circuits

There are three main phases of a virtual call procedure for packet switching:

- Call setup

  — Only required for Switched Virtual Circuit (SVC)

  — Logical channels of the call setup are:

- Incoming only, where the network has ability to initiate calls into a DTE

- Outgoing only, where the DTE has the ability to initiate calls out of channels

- Both-way, where the DTE or the network has the ability to initiate calls out or in

— To minimize call collision:

- Both the DTE and DCE utilize the same channel number simultaneously

- The DTE uses the highest free channel number available to initiate outgoing calls

- The DCE uses the lowest free channel number to initiate incoming calls

▪ Data transfer

▪ Call clearing

The packet structure used in virtual circuits contains several fields. Figure 10.8 lists the packet structure fields in their proper order.

Figure 10.8
Virtual circuit packet structure.

Virtual Circuit Packet Structure
Call Setup Packet
Clear Request and Clear Confirm Packet
Data Packet
Supervisor Packets
Signaling Network Failure Packets
Recovery from a Network Failure

Generic Call Setup Packet

The packet structures discussed do not relate to any specific network protocol, but rather relate to all networks. Any virtual circuit network will have a call setup packet.

Any call setup packet:

- Must contain a virtual circuit number (short identifier to be used in subsequent data packets)

- Must contain a destination address

- May require a source address (not required because the router is aware of which workstation resides at the end of each input line)

- May insert a source address in the call setup, from the router to which the workstation is attached

- Provides a number of facilities; two of the most commonly used facilities allow the user to:

    - Make special requests relative to this call (example: reverse charging)

    - Send long packets (example: if a network is lightly loaded and a user wishes to send a long message, it is more efficient sending it in large blocks)

- Usually contain a small user data field (example: forward a password to the destination so that it may establish its identity)

Generic Clear Request and Clear Confirm Packet

Clear request and clear confirm packets:

- Must have a clear request and a clear confirm packet type to allow the network take down a virtual circuit if desired

- Clear request packet (Figure 10.9) contains:

    - A virtual circuit number associated with this transmission

- A field holding the "reason for the clearing action" may be inserted

- A field normally reading "normal termination"

- An indication that a failure has occurred and that the call could not be continued

- Information useful to the receiving end to determine what subsequent action it should take

Figure 10.9
The Clear Request packet.

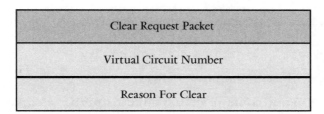

Clear Request Packet

Virtual Circuit Number

Reason For Clear

- Clear confirm packet (Figure 10.10):

    - The beginning virtual circuit number

    - The received original virtual circuit number associated with this transmission, which indicates the virtual circuit has been cleared and that no further billing will take place

Figure 10.10
The Clear Confirm packet.

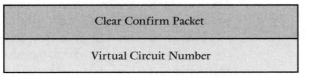

Clear Confirm Packet

Virtual Circuit Number

Generic Data Packet

Components of any data packet (Figure 10.11):

Figure 10.11
A generic data
packet.

Data Packet
Virtual Circuit Number
Sequence Number
Acknowledgment Number
More Packets in Message Indicator
User Data

- Must contain user data

- Must contain virtual circuit number associated with this transmission

- Contain a sequence number field:

  — Used for error control

  — If packets are received out of sequence

    - Indicates that a failure has occurred on the virtual circuit

    - Some action must be taken

  — Controls flow through the network or the user:

    - The network may transmit a Receiver Not Ready (RNR) packet on a specific virtual circuit indicating that it has received a packet, but will not accept subsequent packets on that particular virtual circuit because of traffic congestion. The user must hold packets on this virtual circuit until such time as a Receiver Ready (RR) packet is received, indicating that the congestion has cleared

- Contain an acknowledgment number field

  — Used for error control

Every packet, except for the last packet in the message, contains a single More Packets indicator that specifies that more packets follow. The last packet contains a No More indication, so that reassembly can proceed.

Supervisory Packets

Any supervisory packet:

- Must contain a Receiver Ready (RR) packet (Figure 10.12) which has the following purposes:
  - Sends an acknowledgment when there is no returning data
  - Clears the Receiver Not Ready (RNR) condition

Figure 10.12
Receiver Ready (RR) packet.

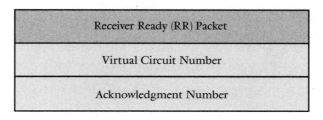

| Receiver Ready (RR) Packet |
| Virtual Circuit Number |
| Acknowledgment Number |

- Must contain Receiver Not Ready (RNR) (Figure 10.13), which:
  - Is used for flow control
  - Is cleared by Receiver Ready (RR)

Figure 10.13
Receiver Not Ready (RNR) packet.

| Receiver Not Ready (RNR) Packet |
| Virtual Circuit Number |
| Acknowledgment Number |

Signaling Network Failure Packets

The components of any signaling network failure packet are:

- Reset packet (Figure 10.14), which:

    — Is used to signal that the circuit is open and available but that some data may have been lost

    — Contains a "Reason for Reset" field that is used to indicate the reason that has been detected

Figure 10.14
Reset packet.

Reset Packet
Reason for Reset

- Restart packet (Figure 10.15), which indicates that:

    — A more serious problem than Reset packet exists

    — A problem has arisen that makes the virtual circuit inoperable

    — The virtual circuit must be reestablished

    — Has a field containing the reason for the generation of the restart packet

Figure 10.15
Restart packet.

Restart Packet
Reason for Restart

Recovery from a Network Failure

The following detailed sequence of events may occur when a workstation has received a Reset on virtual circuit 11 from the network:

- Workstation sends for status information to the receiver to determine how much data had been received prior to the reset event

- The application program has no idea of events occurring in the network

- Workstation places some type of indicator in the packet suggesting that this is a special data packet destined for the Network Layer software at the terminating end of the connection

- The workstation turns on the Q or *Qualified Data bit,* which is data destined for the protocol processor and not the application

- The Qualified Data packet asks the question "Where were we?"

- A returning Qualified Data packet indicates that the receiving protocol layer obtained all packets up to and including number 4 prior to the network failure

- Ordinary data packets may now resume transmission over the virtual circuit

- The data from the previous number 5 packet is sent on its way

- The packet sequence number and acknowledgment fields are reset to 0, so that the routers can reinitialize their counters since they had lost track of sequence numbers prior to the failure

The recovery from a network failure requires agreements between the two workstations. Recovery is an end-to-end protocol, not a network protocol. See Figure 10.16.

Figure 10.16
Recovery from a
network failure.

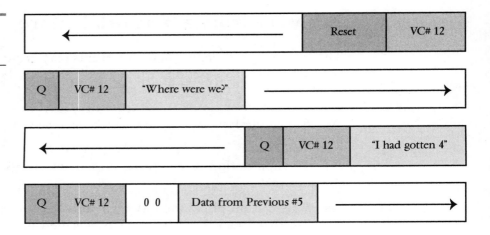

Fast Select Procedure

The fast Select Procedure allows a user to obtain datagram service in a virtual circuit network (where transaction processing is facilitated by datagram, rather than virtual circuit structure). Here are the steps to the Fast Select Procedure:

- If a user wants to send a datagram, the user sends a Call Setup packet and requests that the network provide the facility known as Fast Select

- This request informs the network that the user is sending a call request with a facility request for an expanded data field in the Call Request packet

- Call Request is requesting to place a long user data field at the conclusion of the Call Setup packet

- User then puts the datagram in the data field of the Call Request

  - If recipient accepts, network normally sends a Clear Indication as a response to Fast Select.

    - The Clear Indication may contain user response data

  - If recipient does not accept Call Request, recipient rejects call and puts response in data field of Call Reject

- User sends data across the network and receives a single packet response

- Network routers have no table entries at the conclusion of the exchange

- Exchange appears like a datagram exchange despite the virtual circuit nature of the network

Packet Switching and Public Data Networks

General milestones showing when Packet Switching and Public Data Network came about:

- 1964

    — First described by Paul Baran (Rand Corporation).

- 1967

    — Department of Defense Advanced Research Projects Agency Network (ARPANet) was designed and packet switching was tested as a real strategy for data communications. This design was done under the leadership of Lawrence Roberts.

- 1969

    — Implemented by Bolt Beranek and Newman, the ARPANet was the first network to show the effectiveness of packet switching.

- Late 1960s and early 1970s

    — The Public Data Network (PDN) evolved

    — Similar to a public telephone network, the network acted as a "carrier" for the use of computer data communications.

- Early 1970s

    — ARPANet success proved that packet switching was an efficient, inexpensive, and viable technology for data networks.

Frame Relay

Frame Relay LAN internetworking standards were introduced by the CCITT in 1988 and enhanced in 1990, 1992, and 1993 to provide a fast and efficient method of transmitting information from a user device to LAN bridges and routers. Frame Relay is similar to the X.25 and ISDN network specifications, with additional enhancements. Figure 11.1 shows how Frame Relay fits into the OSI Telecommunications Reference Model.

Figure 11.1
Frame Relay in relation to the OSI Reference Model.

OSI and Frame Relay Comparison	
OSI Reference Model	Frame Relay
Presentation	
Application	
Session	
Tansport	
Network	Frame Relay
Data Link	
Physical	

The technology behind Frame Relay is similar to that of X.25 and ISDN. The advantages of Frame Relay over these services are:

- May transmit a high volume of data

- Has the capability for high bandwidth LANs and WANs

- Uses multiplexing along with virtual circuit techniques

- Interfaces with networks that are capable of doing their own error checking

- Does not incorporate extensive packet-error checking on intermediate nodes

- Achieves a high-speed data transmission form of packet switching

- Transmission speed (data rates) are usually:
    - 56 Kbps and 1.544 Mbps for T1
    - 44.7 Mbps for T3
- Uses fiber optic cable
- Carries smaller packet sizes
- Contains less error checking
- Used with TCP/IP- or IPX-based networks that handle end-to-end error checking at the DTE
- Capable of handling time-delay insensitive traffic, such as LAN internetworking and image transfer
- Uses variable-sized packets (frames)
    - Multiplexed packet-switched packets have variable sizes
    - Frames completely enclose the user packets they transport
    - User packets are enclosed in larger packets (frames) that add addressing and verification information
    - Frames may vary in length up to a design limit of usually 1 kilobyte or more

Frame Relay Structure

Frame Relay structure is based on the LAPD protocol standard. The difference between the Frame Relay structure and the LAPD structure is that the frame header is altered slightly to contain the *Data Link Connection Identifier* (DLCI) and *congestion bits,* in place of the normal address and control fields. Figure 11.2 shows the Frame Relay structure and the packet fields used.

Frame Relay Packet

The Frame Relay packet contains following field formats:

- Flag field
    - Signals the beginning of the frames

- Frame Relay Header field

  — Data Link Connection Identifier (DLCI)

    - Stores the virtual circuit number

    - Each virtual circuit created in Frame Relay is given an ID number to distinguish it from other circuits

  — Command/Response bit (C/R)

    - Indicates whether the packet holds a command or a response type of communication

  — Forward Explicit Congestion Notification (FECN)

    - Upon detection of network congestion, the FECN bit is changed to notify the receiving node

  — Backward Explicit Congestion Notification (BECN)

    - This bit is changed to notify the sending node that there is network congestion

  — Discard Eligibility indicator (DE)

    - When bit is changed, it signals the receiving node to discard packets to relieve network congestion

  — Address Extension bit (AE)

    - This bit shows that extended addressing is used and additional virtual circuits have been created. It is not yet implemented for practical use

- Information field

  — Contains the data for the destination node; size varies per vendor

- Frame-Check Sequence (FCS) field

  — Uses CRC error checking

- Flag field

  — Indicates the end of the frame

Figure 11.2
Frame Relay structure
and packet fields.

Frame Relay Structure									
Flag	Frame Relay Header							Information	FCS	Flag

DLCI	C/R	EA	DLCI	FECN	BECN	DE	EA
8 7 6 5 4 3	2	1	8 7 6 5	4	3	2	1

Explicit Congestion Notification (ECN)

Explicit Congestion Notification (ECN) bits are used to notify user devices of congestion in order to reduce load. If devices are not aware of congestion, they may become congested to the point where they cannot process new data transmissions and begin to discard frames. The discarded frames are retransmitted, which causes more congestion.

There are two bits in the Frame Relay header that signal the user device that congestion is occurring on the line:

- Forward Explicit Congestion Notification (FECN) bit

  — Bit is changed to 1 when congestion occurs during data transmission

  — Frame is transmitted downstream toward the destination location

  — All downstream nodes and the attached user device learn about congestion on the line

- Backward Explicit Congestion Notification (BECN) bit

  — Bit is changed to 1 when congestion occurs during data transmission

  — Frame is traveling back toward the source of data transmission on a path where congestion is occurring

  — The source node is notified to slow down transmission until congestion subsides

Consolidated Link Layer Management (CLLM)

Consolidated Link Layer Management (CLLM) was by defined by the American National Standards Institute (ANSI) to prevent the congestion that may be caused by no frames traveling back to the source node. In this situation, the network will want to send its own message to the problematic source node.

CLLM, service port 1023, is reserved for sending Link Layer control messages from the network to the user device. ANSI standard T1.618 contains a code to identify the cause of congestion. It also contains a listing of all DLCIs that should reduce their data transmission to lower congestion.

Status of Connections

Permanent virtual circuits (PVCs) have a corresponding DLCI that may be needed to transmit information regarding their connection:

- Valid DLCIs for the interface
- Status of each PVC
- Transmitted using the reserved DLCI 1023 or DLCI 0
- *Multicast* status
  - Router sends a frame on a reserved DLCI
  - Reserved DLCI is known as a *multicast group*
  - Network replicates the frame
  - Network delivers the frame to a predefined list of DLCIs
  - The network broadcasts a single frame to a collection of destinations

Discard Eligibility (DE)

Discard Eligibility provides the network with a signal to determine which frames to discard when there is congestion on the line. Frames with a DE value of 1 will be discarded before other frames on the network.

Error Checking

Frame Relay service eliminates time-consuming error-handling processing through the following functions:

- Improved reliability of communication lines
- Increased error-handling sophistication at end stations
- Discarding of erroneous frames

Frame Relay is recognized as a "fast packet" high-speed data transmission that utilizes the newer network technologies that have error checking procedures on intermediate nodes. By utilizing these newer network technologies, Frame Relay does not incorporate extensive error checking. Protocols used with Frame Relay are either TCP/IP- or IPX-based networks protocols. Both these protocols handle end-to-end error checking at the DTE.

Frame Relay does look for the following errors that were not detected:

- Bad frame check sequences
  - Discards bad packets not found by intermediate nodes
- Heavy network traffic
  - Discards packets if it detects heavy network congestion

Multiprotocol over Frame Relay (MPFR)

Multiprotocol over Frame Relay is a method of encapsulating various LAN protocols over Frame Relay. Frames require information to identify the protocol carried within the Protocol Data Unit (PDU). MPFR allows the receiver to properly process the incoming packet. All protocols encapsulate their packets within a Q.922 Annex A frame.

Fields contained in the Q.922 Annex A frame are:

- Q.922 Address
  - 2-octet address field

- — Some networks may contain option 3 or 4 octets
- — Contains the 10-bit DLCI field
- Control
    - — Q.922 control field
    - — UI value is 0×03
    - — Used unless negotiated otherwise
- Pad
    - — Used to align the remainder of the frame to a two-octet boundary
    - — May contain 0 or 1 pad octet within the pad field
    - — Value is always 0
- Network Layer Protocol ID (NLPID)
    - — Administered by ISO and CCITT
    - — Identifies the encapsulated protocol
    - — Identifies the type of data packets as:
        - Routed packets
        - Bridged packets
    - — If protocol does not have an assigned NLPID, the NLPID value indicates the presence of a Sub-Network Access Protocol (SNAP) header
- Frame Check Sequence (FCS)
    - — 2-byte frame check sequence

Figure 11.3 shows the format of the Multiprotocol over Frame Relay frame.

Figure 11.3
Frame format for
Multiprotocol over
Frame Relay.

Multiprotocol over Frame Relay Frame Structure
Q922 Address
Control
Optioal Pad (0X00)
NLPID
. . .
Data
. . .
FCS
Flag (7E Hex)

Figure 11.3 — Frame format for Multiprotocol over Frame Relay.

SNAP Header

A SNAP header is present when a protocol does not have an NLPID already assigned. If the NLPID is not assigned, the default field specifies the SNAP header presence. The SNAP header provides a mechanism to allow easy protocol identification.

The format of the SNAP header is shown in Figure 11.4.

Figure 11.4
SNAP header format.

SNAP Header	
Organizationally Unique Identifier (OUI)	Protocol Identifier (PID)
3 bytes	2 bytes

Figure 11.4 — SNAP header format.

Both the *Organizationally Unique Identifier* (OUI) and the *Protocol Identifier* (PID) identify a distinct protocol.

Frame Relay Data Packet Types

Frame Relay has two basic types of data packets that travel within the Frame Relay network:

- Routed packets
- Bridged packets

These packets have distinct formats and must contain an indicator that the destination uses to interpret the contents of the frame. This indicator is embedded within the NLPID and SNAP header information.

Routed Packets

All devices have the ability to accept and interpret both the NLPID encapsulation and the SNAP header encapsulation for a routed packet. The SNAP header format for a routed packet contains:

- Organizationally Unique Identifier (OUI)
    - Identifies an organization that administers the meaning of the Protocol Identifier (PID)
- Protocol Identifier (PID)
    - Identifies the type of protocol

Exception: If the assigned protocol requires more numbering space than the NLPID provides, a specific NLPID value will not be assigned. When these packets are routed over Frame Relay networks, they are sent using the NLPID 0x80 followed by SNAP. There is one pad octet to align the protocol data on a two-octet boundary.

Bridged Packets

Bridged packets are structured in the NLPID and SNAP header fields as follows:

- Packets are encapsulated using the NLPID value of 0×80, which indicates presence of a SNAP header

- One pad octet aligns the data portion of the encapsulated frame

- SNAP header identifies the format of the bridged packet

  — OUI value used for this encapsulation is the 802.1 organization code 0×00-80-C2

  — PID portion of the SNAP header specifies the form of the Media Access Control (MAC) header

  — PID indicates whether the original FCS is preserved within the bridged frame

Virtual Circuits

A virtual circuit is a logical path from an originating point in the network. Frame Relay frames are transmitted to their destination by way of virtual circuits. By offering virtual circuits, Frame Relay offers advantages to both dedicated lines and X.25 networks for connecting LANs to bridges and routers:

- Virtual circuits

  — Consume bandwidth only when they transport data

  — Multiple virtual circuits can exist simultaneously across a given transmission line

  — Each device can use more bandwidth as required

  — Can operate at higher speeds

Frame Relay uses multiple virtual circuits over a single cable medium. Each virtual circuit constitutes a logical rather than physical connection data path between two communicating nodes.

There are two virtual circuit types within Frame Relay:

- Permanent Virtual Circuit (PVC)

- Switched Virtual Circuit (SVC)

Permanent Virtual Circuit (PVC)

The following characteristics describe the Permanent Virtual Circuit (PVC):

- Administratively set up by the network manager
- Dedicated point-to-point connection
- A continuously available path between two nodes
- Contains a circuit ID
- Remains an open connection for transfer of communication at any time
- One cable medium can support multiple virtual circuits going to various network destinations
- Involves only the Physical Layer and the Data Link Layer of the OSI model
- Physical Layer handles the signaling of transmission
- Data Link Layer handles the virtual circuits

Switched Virtual Circuit (SVC)

The following characteristics describe the Switched Virtual Circuit (SVC):

- Call-by-call set up basis
- Establishment of a transmission session
- A call control signal is distributed between the nodes to connect and disconnect communication
- Allows the user network provider to determine the data through-put rate, which can be adjusted to the needs of the application and network traffic conditions
- Multiple SVCs can be supported on a single cable from point to point

- Newer technology than PVC
- Uses the Physical Layer, Data Link Layer, and the Network Layer of the OSI model:
    - Physical Layer handles the signaling of transmission
    - Data Link Layer handles the virtual circuits
    - Network Layer handles the call control signal protocols

Transport Layer
(Level 4)

Telecommunications service resides primarily in the first three layers of the OSI Reference Model. However, with today's demands for new technology services, protocols have been developed to increase broadband capacity and the speed of data transmission.

Remember, the telecommunications network core is the first three layers of the OSI Reference Model. The fourth layer is the Transport Layer, which contains protocols to operate on top of these layers. Layer 4 protocols allow multiplexing, demultiplexing, faster transmission speeds, and higher bandwidth. As a result, customers may enjoy video-on-demand (VOD), faster Internet connections, and other high speed services.

As you read this chapter, some information will sound very like information already provided for the first three layers of the OSI Reference Model. The Transport Layer is a summation of Layers 1 to 3, with the addition of specialized protocols. Chapter 13 details these specialized protocols that are found in the Transport Layer.

Transport and the OSI Reference Model

The Transport Layer (Figure 12.1) segments and reassembles data into a data stream. The Transport Layer data stream is a logical connection between the endpoints of a network. The upper three Application, Presentation, and Session Layers are concerned with application issues; the lower four layers are concerned with data transport issues.

Figure 12.1
Transport Layer within the OSI Telecommuncations Reference Model.

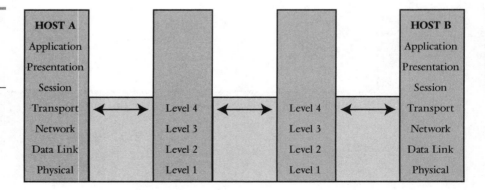

The Transport Layer provides a data transport service that shields the upper layers from transport implementation details. Services provided by the Transport Layer include:

- Segmentation of upper-layer applications

- Reliable transport over an internetwork

- Guarantee that data is transmitted reliably from the point of origin to the destination node

- Reliable service

- Assurance that data is transmitted and received in sequential order

- Mechanisms for the establishment, maintenance, and orderly termination of virtual circuits

- Establishment of a high level of packet error checking

- Handling transport for fault detection and recovery

- Maintenance of information flow control

Transport Layer Classifications

The Transport Layer uses peer protocols from within its layer to employ several reliability measures. These protocols are classified as follows:

- Class 0

  - Simplest protocol

  - Performs no error checking or flow control

- Class 1

  - Monitors for packet transmission errors

  - If an error is detected, requests the sending node's Transport Layer to resend the packet

- Class 2

— Monitors for transmission errors and provides flow control between the Transport Layer and the Session Layer

- Class 3

    — Provides the same functions as Classes 1 and 2

    — Capable of recovering lost packets in certain situations

- Class 4

    — Performs the same functions as Class 3

    — More extensive error monitoring and recovery

Transport Layer Flow Control

Flow control ensures the integrity of data. Two causes of data congestion are:

- High-speed computer may generate traffic faster than a network can transfer it

- Multiple computers simultaneously send datagrams to a single destination

    — Data is temporarily stored in memory until the system can process the information

    — If traffic continues, the system exhausts its memory and discards additional datagrams upon arrival

To prevent lost data from congestion:

- Transport Layer issues a Not Ready indicator to the transmitter to stop sending data

- When buffer is clear and ready for additional data, the receiver sends a Ready transport indicator

This method prevents the receiving host from overflowing its buffers and encountering lost data.

Transport Layer Connection Procedures

The Transport Layer maintains a connection-oriented relationship between the communicating end systems. Four procedures are followed by the Transport Layer to ensure reliability:

- Delivery of segments is acknowledged to the transmitter
- Retransmission of segments is not acknowledged
- Reordering of segments into their correct sequential order at the destination point
- Congestion control

Multiple applications can share the same transport connection. Transport functionality is accomplished segment by segment. Different applications can send data segments on a first-come, first-served basis. These segments can be intended for the same destination or for many different destinations.

To transmit data from the Transport Layer of one system to the Transport Layer of another system, the following steps must occur:

- Before transmission of data begins, the device sets a port number for each software application
- Additional bits are included to encode:
    - Message type
    - Originating program
    - Protocols used
- Upon receipt, the destination device separates and sorts the segments in order to pass the data to the correct destination application, which is determined by the assigned port number
- The destination system establishes a connection-oriented session with its peer system
- Both transmitting and receiving systems notify their operating systems that a connection will be initiated
- One system transmits an acknowledgment that must be accepted by the receiving system

- Once synchronization has been accomplished, a connection is established, and data is transferred

- Both systems continually communicate for verification of correct data

Figure 12.2 illustrates the basics of a telecommunications transmission in which the Transport Layer is used as a foundation. Chapter 13 provides information for the most popular protocols used in the Transport Layer.

Figure 12.2
Transmitter and receiver establish connection in the Transport Layer.

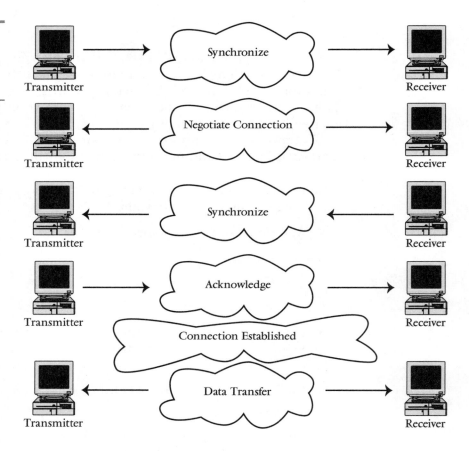

Integrated Services Digital Network (ISDN)

The Integrated Services Digital Network (ISDN) is a protocol that describes how users of a private or public network exchange information about incoming and outgoing calls.

ISDN was introduced in the 1970s, made official in 1984, and later refined in 1988 by the CCITT. The CCITT recommendations define a standard process to allow an ISDN connection to any location in the world. ISDN carries all information in an end-to-end digital network—no analog transmission services are used. As with other digital protocols, there must be a set of standards to allow digital networks to interconnect to analog networks.

The benefits of ISDN are:

- Layered protocol structure compatible with OSI

- Provides voice, data, and video services over one network

- Network management services are offered via intelligent nodes

- Communication channels are offered in multiples of 64 Kbps, such as 384 Kbps and 1536 Kbps throughput

- Provides switched and nonswitched connection services

- May provide videoconferencing through high-bandwidth capabilities

ISDN incorporates the Physical, Data-Link, Network, and Transport Layers of the OSI model. Similar to X.25, it uses LAPB and the data-link layer to ensure the maximum detection of communication errors. Figure 13.1 shows ISDN in relation to the OSI-Telecommmunications Reference Model.

ISDN is designed to be compatible with many existing digital networks, such as ATM, X.25, and T1 (T1 has a data rate of 1.54 Mbps). ISDN is divided into 64-Kbps channels. Table 13.1 illustrates the channels of an ISDN line.

Figure 13.1
ISDN and the OSI
Reference Model.

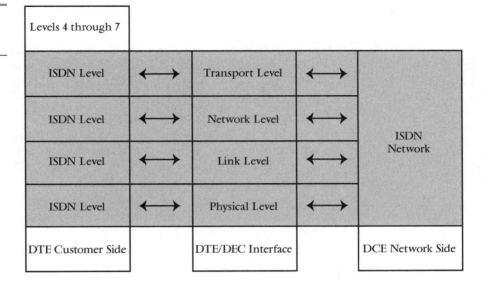

Levels 4 through 7			
ISDN Level	←→ Transport Level ←→		
ISDN Level	←→ Network Level ←→		ISDN Network
ISDN Level	←→ Link Level ←→		
ISDN Level	←→ Physical Level ←→		
DTE Customer Side	DTE/DEC Interface		DCE Network Side

TABLE 13.1

Channels of an
ISDN Line

Information Rate	Channel	Applications
64 Kbps	B	8 KHz General Purpose Communications
64 Kbps	B	8 KHz Digitized Speech Number identification Multi-party calling Call completion
64 Kbps	B	3.1 KHz Audio Text FAX Combination text and FAX
64 Kbps	B	8 KHz Alternate Transfer of Speech
384 Kbps	B	8 KHz Video and PBX Link Fast FAX CAD/CAM Imaging High-Speed Data LAN Internetworking
1536 Kbps	H11	Same services as 384 Kpbs

ISDN Integrated Services

The ISDN protocol's signaling messages are packets. The ISDN protocol provides a wide range of services for voice, nonvoice, data, audio, video, graphics, digital services, and interactive data transmissions.

All protocol services are available at a common point using a *universal socket*, which comprises both hardware and software. A universal socket is located between the user and the network and provides a single connection point where the user receives services. The universal socket has a procedure for requesting services.

Information can be carried over the local loop in a digital manner. Time Division Multiplexing (TDM) of a service (voice, data, video, etc.) facility can be simultaneously provided over the same facility.

The Central Office switch can support voice, video, and audio because these services use a circuit mode service. Packet mode data may require the presence of a second switch in the Central Office. This approach is taken with an integrated circuit-packet switch.

ISDN Public Network

Integrated Services Digital Network (ISDN) is viewed as a standard for connection to a public network (Figure 13.2). By this method, exchange of signaling information occurs between users and the network. This connection may be from a user to a local exchange carrier or directly to an Internet carrier. Equipment behind the user's ISDN connection point may support a variety of standards. ISDN public network requires Signaling System 7 (SS7) support.

Figure 13.2
ISDN public network.

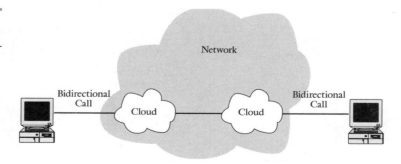

Network

Bidirectional Call

Cloud

Cloud

Bidirectional Call

ISDN Private Network

Integrated Services Digital Network (ISDN) is a protocol that may be used in a private network (Figure 13.3). An ISDN private network consists of a connection of two or more switches with ISDN signaling between them. Local exchange carrier and Internet carrier transmission facilities can be used to provide the point-to-point transmission. The private network does not require SS7.

Figure 13.3
ISDN private
network.

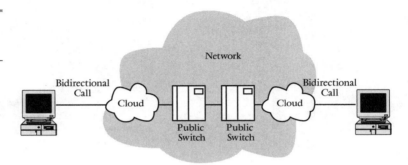

Interface from User to ISDN

Before ISDN, each service required a different interface: To connect to an X.25 packet-switched service required a separate circuit, and any changes in service would take 30 days or more. Using ISDN, all services are obtained at a single, standard physical interface. All services are available when the equipment is plugged into this single socket. Control of service selection is handled by the customers, and services can be obtained or modified whenever the service is required. Figure 13.4 illustrates the single-service point concept of ISDN.

ISDN Services Data Rates

ISDN service data rates vary based on the signal being transmitted.

Figure 13.4
A single service point
providers numerous
ISDN services.

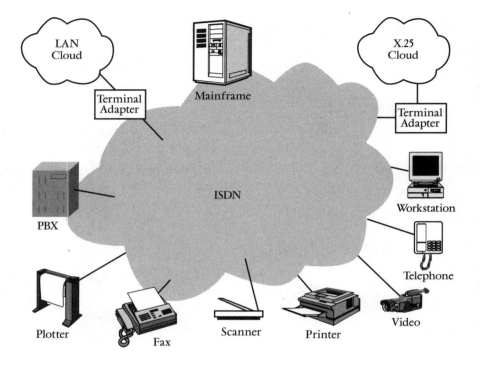

- Human speech

  - Requires a circuit mode facility because voice transmission produces a long hold time, delay-sensitive signal

  - Network uses *digital circuit multiplication,* which recognizes the pauses in speech

- Nonvoice, audio data

  - Utilizes a circuit mode-type facility

  - Network does not use digital circuit multiplication because it need not allow for pauses (i.e., music is continual where speech contains pauses)

- Interactive data

  - Utilizes a packet mode facility

  - Bit rate is low because human beings are not capable of sending or receiving data at high rates

— Interfaces such as RS-232-C limit the bit rate of terminal devices to about 19.2 Kbps

- Bulk data

   — Requires high-speed circuit mode transmission

   — Bulk transport applications transmit sufficiently on T1 or T3

- Video

   — Supports non-full motion transmission only (not streaming, live video)

   — Quality depends on the amount of motion present and the compression technique applied (e.g., the transmission of a still picture comes through with adequate quality, but a motion picture would turn out unsatisfactorily)

Table 13.2 shows the comparative data rates for ISDN transmissions.

TABLE 13.2

ISDN Data Rates

Service	Data Rate	Facility Type
Voice	8, 16, 32, 64	Circuit
Nonvoice audio	8, 16, 32, 64	Circuit
Interactive data	2.4 to 19.2	Packet
Bulk data	Up to 1536	Circuit
Video	64 to 1536	Circuit

In- and Out-of-Band Signaling

There are two different types of ISDN network signaling:

- In-band signaling
- Out-of-band signaling

In-Band Signaling

Switches must send signaling information internally to provide required services (e.g., testing of trunks, seizing, number requests, etc.). Requests for service and messages that define the status of devices at the opposite ends of the network are all sent through the network. Specific frequencies within the voice band were used to carry signaling information: These signals were said to be sent "in-band." The signal path and the transmission path for the voice were the same.

In-band signaling (Figure 13.5) has a number of drawbacks, the most important of which is that is wasteful of equipment. The entire end-to-end path must be set up to send signals.

Figure 13.5
In-band signaling.

Local CO	Local Toll	Remote CO
	Seize Trunk →	
	Testing Trunk	
	←	Send Number
	Number →	
	←	Alerting
←	Alerting	
	←	Connection
→	Discount	→
		Disconnection →
Release Equipment		

Out-of-Band Signaling

Out-of-band signaling is an alternate choice to in-band signaling the uses a signaling path distinct from the voice path. There are two different signaling out-of-band signaling techniques:

- *Associated signaling* (Figure 13.6)

Figure 13.6
Out-of-band
associated signaling.

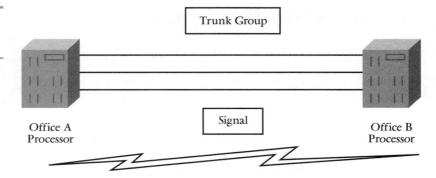

— Signaling path is associated with individual trunk groups

— Signaling information on a path can only be used for the associated trunk groups

■ *Disassociated signaling* (Figure 13.7)

— Independent signaling network is separate from the voice network

— Disassociated signaling systems (CCIS and Signaling System 7) use separate packet-switched networks

Figure 13.7
Out-of-band
disassociated
signaling.

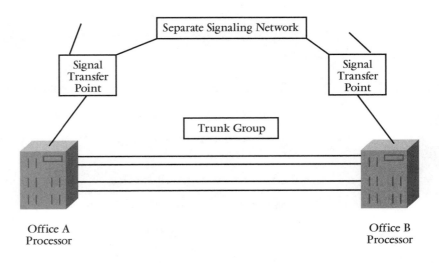

ISDN Data Rate Interfaces

Two common interfaces for ISDN data rates are:

- Basic Rate Interface (BRI)
- Primary Rate Interface (PRI)

Basic Rate Interface (BRI)

Characteristics of the Basic Rate Interface (BRI) include:

- Aggregate data rate of 144 Kbps
- Bit transfer rate of 192 Kbps
- Difference between aggregate data rate and bit transfer rate is overhead associated with signaling, timing, and framing functions
- Point-to-point is supported
- Uses two channels, B and D

B Channel

The characteristics of a bearer on B channel on the Basic Rate Interface are:

- May be point-to-point, point-to-multipoint, or broadcast
- Not shared between devices on the multipoint configuration
- Assigned to a specific device on the multipoint link for as long as that device chooses to use it
- Only source of circuit service
- Provide a packet mode connection
    - A high throughput device might request a B channel for packet exchanges

- — Currently, X.25 packet-switched service is supported on the B channels
- — At current level of ISDN implementation, a packet-switched B channel must be assigned by a service order
- Carries only user information
- Because the Data or D channel carries all service requests, B channels are "clear" for all 64 Kbps to be used for data
- No requirement for special escape sequence is used to indicate that services may be terminated
- Services may be assigned according to:
  - — Demand
    - Services may be required on a demand basis
  - — Reserved
    - Demand services can be viewed as a reserved dialed service in the phone network
  - — Permanent
    - Permanent services work as leased line services
- Any pattern of information is acceptable
- Connections may be:
  - — Symmetric mode transmission, where transmission rate is the same for both directions
  - — Asymmetric mode transmission, where rates are not the same in both directions
  - — Unidirectional mode transmission, where transmission occurs in one direction only
- Set up:
  - — Must be setup via a D channel exchange
  - — D channel setup requests (i.e., channel identifier, speed/throughput, etc.)
  - — Local Exchange (LE) returns Call Processing or a request for more information (Setup Acknowledged)

— Remote Local Exchange (LE) sends setup to destination(s)

— When destination(s) issue a Connect, remote Local Exchange (LE) notifies local Local Exchange (LE)

— Local Exchange (LE) sends Connect or Connect Reject (cause)

D CHANNEL.   The characteristics of the Data or D channel on the Basic Rate Interface are:

- Shared between all devices that hang on the multipoint configuration

- Used to obtain a B channel

- Operates in a Time Division Multiplexed (TDM) fashion with the B channels

- Carries all signaling information

- Operates in packet mode only

- May be used to send packet data

- Cannot support circuit mode connections

- Signaling on the D channel relates only to B channel at the same interface

Figure 13.8 shows the layered protocols used on the D channel.

Primary Rate Interface (PRI)

The characteristics of the Primary Rate Interface include:

- Supports high-speed networks with data rates up to 622 Mbps

- Point-to-point only

- For large computers

- May be configured as either 23 B channels plus one D channel at 64,000 bits per second or as 24 B channels

- Uses two channels: B and D

Figure 13.8
Layered protocol on
the D channel.

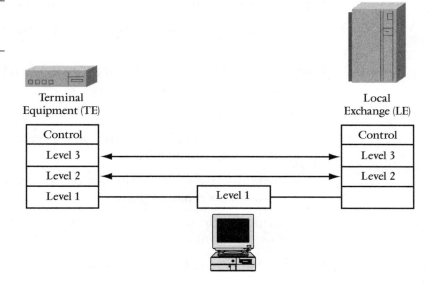

B CHANNEL. The characteristics of the B channel on the Primary
Rate Interface are:

- Not shared between devices on the multipoint configuration

- Assigned to a specific device on the multipoint link for as long as
that device chooses to use it

- The only source of circuit service

- Provides packet mode connection

  — A high throughput device might request a B channel for
  packet exchanges

  — Currently, X.25 packet-switched service is supported on the
  B channels

  — At current level of ISDN implementation, a packet-switched
  B channel must be assigned by a service order

- If an application requires more bandwidth than is available on
a single B channel, ISDN supports of grouping of multiple B
channels

- H0 channels are six contiguous B channels (384 Kbps)

- May have three or four H0 channels, depending configuration set up

- If bit rate of an H0 channel is not adequate, H11 or H12 channels can be used

- Connections may be:

  — Symmetric mode transmission, where transmission rate is the same for both directions

  — Asymmetric mode transmission, where rates are not the same in both bidirections

  — Unidirectional mode transmission, where transmission occurs in one direction only

- B Channel services may be assigned according to

  — Demand

    - Services may be required on a demand basis

  — Reserved

    - Demand services can be viewed as reserved dialed services in the phone network

  — Permanent

    - Permanent services work as leased line services

- Set up

  — Must be set up via a D channel exchange

  — D channel setup requests (i.e., channel identifier, speed/throughput, etc.)

  — Local Exchange (LE) returns Call Processing or a request for more information (Setup Acknowledged)

  — Remote Local Exchange (LE) sends setup to destination(s)

  — When destination(s) issue Connect, remote Local Exchange (LE) notifies local Local Exchange (LE)

  — Local Exchange (LE) sends Connect or Connect Reject (cause)

D CHANNEL. The characteristics of the D channel on a Primary Rate Interface are:

- Shared between all devices that hang on the multipoint configuration

- Used for all service requests

- Used to obtain a B channel

- Must be used in packet mode

- May be used to send packet data

- Cannot support circuit mode connections

- Two versions

   — United States version

      - 23 B channels and 1 D channel or 24 B channels

      - B channels are 64 Kbps

      - D channel (if present) is 64 Kbps

      - Data transfer rate is 1.536 Mbps

      - Bit transfer rate is 1.544 Mbps

      - H11 supports all 24 B channels used for user data and provides a rate of 1,536,000 bits per second

   — European version

      - 30 B channels and 1 D channel or 31 B channels

      - B and D channels are same as in the United States

      - Data transfer rate is 1.984 Mbps

      - Bit transfer rate is 2.048 Mbps

      - H12 channel supports the rate of 30 B channels

Table 13.3 shows the most commonly used BRI channels. Table 13.4 illustrates the layered protocols used in the PRI channels.

TABLE 13.3

*Most Commonly
Used BRI Channels*

Information Rate	Channel	Applications
64 Kbps	B	8 KHz general purpose communcations Carry data or voice Clear
64 Kbps	B	8 KHz digitized speech Carry data or voice Clear Number identification Multiparty calling Call completion
16 Kbps	D	16 Kbps packet mode Carries signaling packets Carries user data packets

TABLE 13.4

*Most Commonly
Used PRI Channels*

Information Rate	Channel	Applications
64 Kbps	B	31 KHz audio Text Fax Combiation text and fax
64 Kbps	B	8 KHz alternate transfer of speech
384 Kbps	B	8 KHz video and PBX link Fast fax CAD/CAM imaging High-speed data LAN internetworking
1536 Kbps	H	Same services as 384 Kpbs

Upon an ISDN channel connection, a couple of distinctions need to be stressed:

- Implicit

  — Service request at your own interface

- Explicit

  — Service request at another interface

  — Must identify the specific interface at which the service is required

Protocols for the D and B Channels

The ISDN protocol on the Data or D channel is a three-layer protocol modeled after the OSI Reference Model.

- Physical Layer (Level 1)

  — Bits transmitted from the Terminal Equipment (TE) to the Network Termination 1 (NT1)

  — All information physically passes through the NT1

  — Network Termination 1 (NT1) implements Level 1 and provides transmission line termination

- Data Link Layer (Level 2) and Network Layer (Level 3)—General

  — Assumption of peer-to-peer communication between the device at the user site (TE) and the Local Exchange (LE) at the network site

  — Information is interpreted by the Local Exchange (LE)

  — Link between Network Termination 1 (NT1) and Local Exchange (LE) may be TCM or echo cancellation

  — Transparent to the user

- Control function above Level 3 is not specified in ISDN

On the B channel, ISDN implements a single-layer interface.

B Channel Increased Rate Adaptation

ISDN recommendations allow equipment to operate at speeds other than 64 Kbps on the B channel. The adaptation steps are as follows:

- A Terminal Equipment (TE) device operating at a rate slower than 64 Kbps connects to a Terminal Adapter (TA)

- The Terminal Adapter (TA) takes this rate and adapts it upward

- The Terminal Adapter (TA) uses one of a series of recommendations from CCITT to achieve a speed of 61 Kbps

- Even though the equipment is not operating at 64 Kbps, each B channel still transmits at 64 Kbps.

The rate adaptation process may occur in two stages, which are illustrated in Table 13.5.

TABLE 13.5

Two-stage B Channel Increased Rate Adaptation

Initial Rate	Stage 1	Stage 2
9.6 Kbps (not a multiple of 8 Kbps)	16 Kbps (a multiple of 8 Kbps)	64 Kbps
19.2 Kbps (not a multiple of 8 Kbps)	32 Kbps (a multiple of 8 Kbps)	64 Kbps
48.0 Kbps (not a multiple of 8 Kbps)		64 Kbps
56.0 Kbps (not a multiple of 8 Kbps)		64 Kbps

Adaptation of stages is described in Recommendations 1.460, 1.461, 1.463

D Channel Bit Framing

The two layer of the OSI Reference Model that are responsible for bit framing are the:

- Data Link Layer (Level 2)

  — Ensures absence of bit errors or missing messages

  — Framing support

  — I.440 and I.441 (Q.920/Q.921) recommendations specify the operation of the D channel for basic rate interface and primary rate interface

- Network Layer (Level 3)

  — Provides mechanism for obtaining service

  — Messaging support

  — The CCITT Recommendation 1.451 (Q. 931) describes the syntax and semantics of the signaling messages sent between the user and the network

In general, the format of the message is consistent with that of a bit-oriented protocols. The proper order of a frame structure is:

- Flag
- Address
- Control field
- Q.931 signaling message or a user data packet
- Frame check sequence
- Terminating flag

The I.440 and I.441 specifications allow the upper-level layers to assume that the Physical Layer is error-free. Figure 13.9 illustrates bit framing in the D channel.

Bearer Service

A *bearer service* is responsible for the addition of information attributes to signaling data so that the network will know what to do with the packets within the channels. A bearer service consists of two attributes:

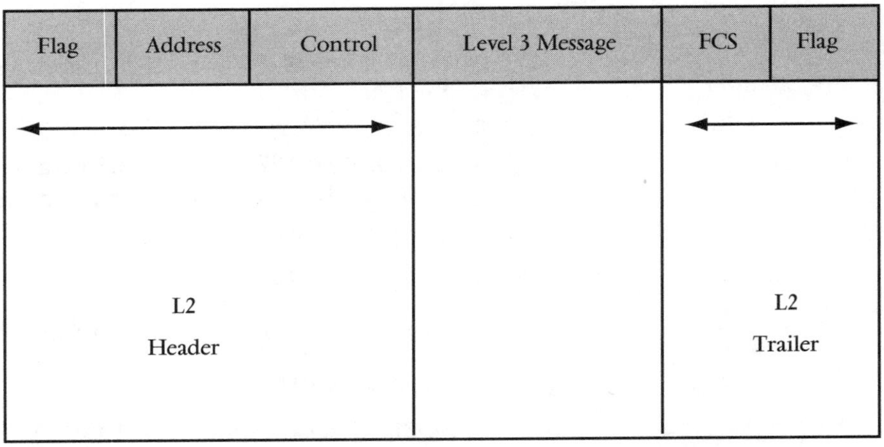

FIGURE 13.9
Bit framing in the D
channel.

Flag	Address	Control	Level 3 Message	FCS	Flag

L2
Header

L2
Trailer

- Information Transfer attributes
  - Important from end-to-end
  - Contain information on services required on receiving end
  - Examples: Circuit service, 64 Kbps, voice mode
- Access attributes
  - Important from user to local exchange carrier
  - Establish agreement between the remote local exchange and the remote user about which band on a channel to use
  - Have no channel requirements to be the same end-to-end
  - Are of local significance only

Teleservices

Teleservice is the combination of a bearer service and another value-added service (e.g., home banking, etc.). Teleservices are critical to the success of ISDN. Figure 13.10 shows some common ISDN teleservices.

Figure 13.10
Teleservices using
ISDN.

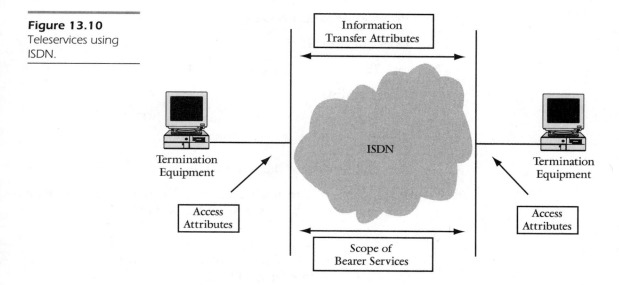

ISDN Data Rate Topologies

As previously discussed, there are two common interfaces for ISDN Data Rates:

- Basic Rate Interface (BRI)
- Primary Rate Interface (PRI)

These interfaces have specific speeds or data rates, and also specific topologies.

ISDN Topology Terminology

- Network Termination Equipment 1 (NT1)
  — Customer premises equipment
  — Provided by the customer
  — Considered the end point of the network
  — Fundamentally a DCE

- — Provides signal conversion

- — Does not provide processing

- Network Termination Equipment 2 (NT2)

  - — A service distributor

  - — A host computer or PBX can serve as a NT2

  - — Partitions ISDN services

  - — Provides services to the attached devices

  - — ISDN devices communicate to a Network Termination Equipment 2 via an unspecified ISDN standard

- Non-ISDN Standard

  - — An interface between Network Termination Equipment (NT1) and Local Exchange Carrier

- Not an ISDN standard

  - — An interface standard specified in the United States to promote multiple suppliers for Network Termination Equipment development

- ISDN-Defined Interface

  - — ISDN standards apply

  - — User terminal equipment must adhere to the interface specification to obtain ISDN services

Basic Rate Interface (BRI) Topology

The Basic Rate Interface (BRI) in ISDN is a point-to-multipoint topology (see Figure 13.11). In a point-to-multipoint configuration:

- D channel must connect to each individual Terminal Equipment (TE) device

- D channel may not be owned by a particular Terminal Equipment (TE) device

- D channel is a Local Area Network (LAN)

- B channel is not shared

- B channel is owned by one particular device at a time

- B channel may operate in circuit- or packet-mode transmission

Figure 13.11
ISDN Basic Rate
Interface (BRI)
topology.

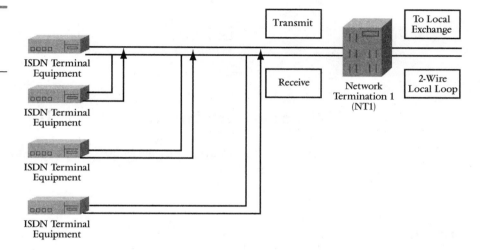

Point-to-multipoint operation has the potential for collision for the following reasons:

- D signaling channel is not assigned permanently to an individual Terminal Equipment (TE) device

- Two Terminal Equipment (TE) devices may decide to transmit at the same time

Because of this potential problem of collision, there must be some type of access control procedure established to prevent one Terminal Equipment (TE) device from transmitting and destroying information from another Terminal Equipment (TE) device.

Figure 13.12 shows an example of an ISDN Basic Rate Interface network diagram flow.

Figure 13.12
ISDN Basic Data Rate
network flow.

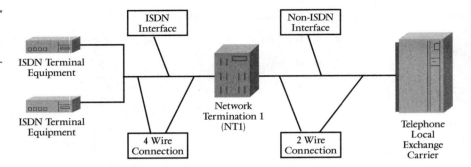

Primary Rate Interface (PRI) Topology

Figure 13.13 shows an example of an ISDN Basic Rate Interface network diagram flow.

Figure 13.13
ISDN Primary Data
Rate network flow.

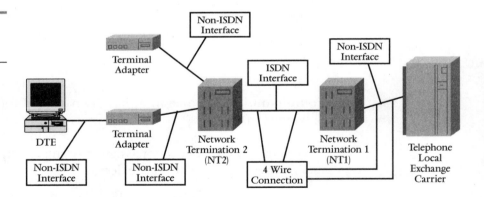

Packet-Mode Data Transport

There are two ISDN options for providing packet-mode data transport:

- Access to packet-switched services
- ISDN virtual-circuit bearer services

Access to Packet-Switched Services

The ISDN does not know how to handle user data packets; therefore, the ISDN must assign a B channel in circuit-mode and provide a connection between the user's packet mode device and a port at a packet switch.

The ISDN does not provide packet services—it only provides a conduit for information flow. The D channel may not be used to send packets in this mode. This service is illustrated in Figure 13.14.

Figure 13.14
Access to packet-switched services.

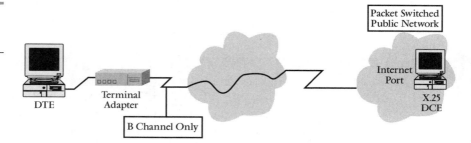

ISDN Virtual-Circuit Bearer Services

When using virtual-circuit bearer services, the ISDN network and the packet network are merged so that the local exchange carrier knows what to do with the user data packets. There is no requirement for the user to establish a circuit to get packets delivered. Virtual-circuit service has the advantage of being able to use the D channel to send small amounts of data, in addition to the B channel for the larger packets. Figure 13.15 shows the virtual-circuit bearer service.

In virtual-circuit bearer services the B channel:

- Carries only user information

- Is "clear" for all 64 Kbps to be used for data, since D channel carries all service requests

- Has no requirement for special escape sequences to be used to indicate that services may be terminated

- Accepts any pattern of information

Figure 13.15
Virtual-circuit bearer
services.

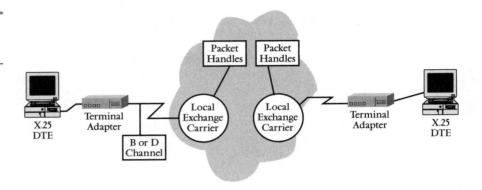

Local Loop Requirements

It is a requirement for providing an integrated service that the local loop carry both voice and data simultaneously. Because of the manner in which some local loops are provisioned, integration of both voice and data may not be possible.

The original requirement for local loops was to provide voice service only. When only concerned with voice services, these services could be provisioned with a variety of features that made the transmission of frequencies higher than those required for voice impossible. *Loading coils* (the addition of repeaters, bridges, routers, etc.) permitted the transmission of voice frequencies over smaller gauge wires (less expensive) at greater distances, with an acceptable attenuation level. However, the loading coil method blocked signals with frequencies above the voice band.

For a local loop to have the capability to transmit both voice and data simultaneously, it is a requirement the transmission line be unloaded. Unloaded coils are straight copper lines with no addition of repeaters, bridges, routers, or other such devices.

Figure 13.16 illustrates the level of acceptance required to transmit voice and data simultaneously. The distance requirement is 14,000 to 18,000 feet, 3300 HZ, and unloaded.

Figure 13.16
ISDN local loop
requirements.

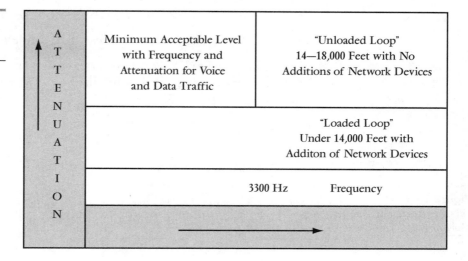

Echo

A two-wire path is a half-duplex medium that transmits signals from both ends simultaneously. This method of transmission makes it difficult to distinguish near-end from far-end signals. A solution is to insert a negative image of the near-end signal into the receiver to cancel the near-end effect and permit the receiver to "hear" only the far-end signals. The use of this technique would work well if it were not for the problem of echo.

The signal may echo from points in the signal path. Gauge changes, bridged taps, and hybrid coils (2-to-4-wire converters) may echo a signal, mimicking the appearance of a far-end signal. There are two solutions:

- Time Compression Multiplexing (TCM)
- Echo cancellation

Time Compression Multiplexing (TCM)

The characteristics of Time Compression Multiplexing (TCM) (Figure 13.17) are:

- Loop is run half-duplex at twice the announced data rate

- The "Ping Pong" methods emulates a full-duplex path

- 16 to 24 bit blocks (288 Kbps) of data are sent in alternating digital bursts and pauses: a burst in one direction—then a pause—then a burst in the opposite direction

- Managed by a central timing control

Figure 13.17
Time Compression
Multiplexing (TCM).

Data Rate n (FDX)	Buffer	Data Rate 2n (HDX)	Buffer	Data Rate n (FDX)
Customer Premise		"Ping Pong"		Local Office

Echo Cancellation

The characteristics of echo cancellation (Figure 13.18) are:

- Processor "learns" the echo characteristic of the loop

- Processor inserts negative images of the echoes into the receiver at precisely the right time to eliminate the effect

- Requires sophisticated circuitry

- Uses a device called a "hybrid," which connects the transmitter and the receiver to the subscription line

- Transmits data in two direction simultaneously, which can cause echo of the transmitted signal

- Determines the amplitude of the echoed signals and subtracts the amplitude from the incoming signals

- Popular approach in the United States

- Also employed by the 5ESS switch

Figure 13.18
Echo cancellation.

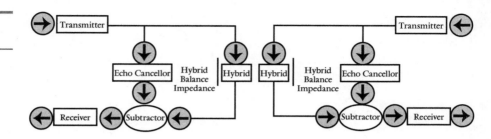

Asynchronous Transfer Mode (ATM)

Asynchronous Transfer Mode (ATM) is an international standard developed by CCITT. The existence of multiple communication standards, such as FDDI, Ethernet, and Token Ring has directed the requirement for an international standard—Asynchronous Transfer Mode (ATM) is such a standard and has gained wide acceptance for network interoperability.

ATM Advantages

ATM has acquired acceptance for telecommunication performance because of five main factors:

- Handles data, voice, and video transmissions
- Dependable and flexible at geographic distances
- Accommodates high-speed telecommunication
- Provides potentially significant cost saving in network resources
- Encodes in fixed-length, 53-byte relay units of data called *cells*

Organizations that have a large investment in client/server technology are quickly moving to implement ATM. New demands to transmit voice and video data, as well as large database queries, require the bandwidth capabilities of ATM.

LAN and WAN Communications

ATM is used for LAN and WAN communications, because it provides flexibility over geographic distances. Connectivity between local, city, and worldwide networks is simplified if users implement a single networking system, and ATM has become popular in the networking industry because it can handle transmission speeds in the gigabyte range. The speed of the ATM technology offers greater flexibility as more organizations push network data throughput with multimedia and client/server applications.

Cost Savings

Cost must always be part of the decision process when determining a transmission method for telecommunication. Many networks must use separate transmission media for voice, video, and data, because the transmission characteristics are different for each service. ATM can handle voice, video, and data on a single network medium. Voice and video transmissions are continuous streams of signals along the cable, and video signals can occupy large bandwidths.

ATM can handle transmission speeds in the gigabyte range. Programming techniques consume more and more bandwidth; because ATM can handle voice, video, and data on a single network medium, it represents a large cost savings in bandwidth and network resources. Because ATM connectivity has the flexibility to be used for both Local Area Networks (LANs) and Wide Area Networks (WANs), it eliminates the need for separate local and wide area distance networks.

ATM Cell Structure

ATM uses a cell-switching technology. Each ATM packet is referred to as a cell. ATM cells have a fixed length of 53 bytes, which allows for very fast switching. ATM creates pathways called *virtual circuits* between end nodes.

The fixed-length ATM cell contains two primary sections (see Figure 14.1):

- Header
- Information

Figure 14.1
ATM cell structure.

ATM Cell Structure
Header—5 octets or 40 bits
Information—48 octets or 384 bits

Information Field Cell Structure

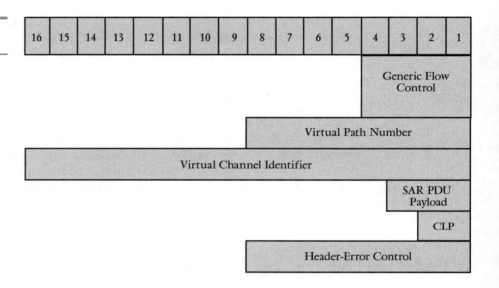

Figure 14.2
Header structure.

The Header contains:

- Flow control information
    - Generic Flow Control (GFC) is included in the field
    - 4 bits long
- Virtual Path Identifier (VPI)
    - Management information for the Physical Layer about the communication channel in use
    - Address information is referenced to the network equipment that the cell travels through to its destination
    - 8 bits long
- Virtual Channel Identifier (VCI)
    - Management information for the Physical Layer about the communication channel in use

- — Address information is referenced to the network equipment that the cell travels through to its destination

- — 16 bits long

- — Payload type (PT)

- — Shows whether the cell payload contains user information or connection management information

- — Indicates whether or not the cell encountered network congestion during transmission

- — 3 bits long

- Cell Loss Priority (CLP)

  - — Indicates whether or not the cell should be transmitted by network equipment when there is high network traffic

  - — A 0 cell has a high priority

  - — A 1 cell can be dropped if there is network congestion

  - — 1 bits long

- Header-Error Control (HEC)

  - — Contains information to indicate if an error has occurred during the transmission of the packet

  - — 8 bits long

ATM is a multiple layered telecommunication system; it operates over the following layers:

- Physical Layer

  - — Conducts the relay cell as a signal

  - — Consists of the electrical transport interface that conducts the cell as a signal

  - — May be transported over coaxial, twisted-pair, or fiber optic cable

- ATM layer

  - — Assembles the cell header

  - — Adds it to the payload data

- Adaptation layer (AAL)

  — Constructs voice, video, and data into the cell payload

ATM Adaptation Layers (AAL)

The ATM Adaptation Layer (AAL) is located in the Segmentation and Reassembly Sublayer (SAR) or Payload field of the ATM cell packet structure.

AAL1 PDU

The AAL1 PDU field (Table 14.1) consists of:

- Sequence Numbers (SN)

  — Numbers assigned to SAR PDU

  — Contains the following fields:

    - Convergence Sublayer Indicator (CSI), which is a residual time stamp for clocking

    - Sequence Count (SC), which is a number for the PDU and is generated by the Convergence Sublayer

- Sequence Number Protection (SNP)

  — Contains of the following fields:

    - Cyclic Redundancy Check (CRC), which is calculated over the SAR PDU Payload header

    - Even Parity Check (EPC), which is calculated over the Cyclic Redundancy Check (CRC) header

- SAR PDU Payload

  — 48-byte user information field

TABLE 14.1

AAL1 PDU Cell Structure

Sequence Number (SN)		Sequence Number Protection (SNP)		
Convergence Sublayer Indicator (CSI)	Sequence Count (SC)	Cyclic Redundancy Check (CRC)	Even Parity Check (EPC)	SAR PDU Payload
1 bit	3 bits	3 bits	1 bit	47 bytes

AAL2

In delay-sensitive applications, the AAL2 service (see Table 14.2) provides short, low-rate, and variable packets that provide efficient bandwidth transmission. AAL2 also has the capability to provide for variable payload within and across cells. The structure of the AAL2 packet is:

- Service Specific Convergence Sublayer (SSCS)
    - Channel Identification (CI) valves:
        - 0 Not Used
        - 1 Reserved layer management peer-to-peer procedures
        - 2—7 Reserved
        - 8—255 Identifies AAL2 user (248 total channels)
    - Length Indicator for each individual user's packet payload length:
        - 1 less the packet payload
        - Default value of 45 bytes
        - May be set to 64 bytes
    - User-to-User Indication (UUI) to provides link between the CPS and SSCS that satisfies the higher layer application:
        - 0—27 Identification of SSCS field entries

- - 28, 29 Reserved for future standardization
  - - 30, 31 Reserved for layer management (OAM)
  - — Header Error Control (HEC)
- ■ Common Part Sublayer (CPS)
  - — Offset Field (OSS)
    - - Identifies start location of the next CPS packet with the CPS-PDU
  - — Sequence Number (SN)
    - - Protects data integrity
  - — Parity (P)
    - - Protects start field from errors
  - — SAR PDU payload
    - - SAR PDU Information field
  - — Padding (PAD)

TABLE 14.2   The AAL2 Cell Structure

Service Specific Convergence Sublayer (SSCS)				Common Part Sublayer (CPS)				
Channel Identification (CID)	Length Indicator (LI)	User-to-User Indication (UUI)	Header Error Control (HEC)	Offset Field (OSF)	Sequence Number (SN)	Parity (P)	SAR PDU Payload	Padding (PAD)
8 bits	6 bits	5 bits	5 bits	6 bits	1 bit	1 bit		0—47 bytes

AAL3/4

AAL3/4 functions:

- Support message and streaming modes
- Provide ATM point-to-point and point-to-multipoint connections
- Support a connectionless Network Layer (Class D)
- Support Frame Relay telecommunication service in Class C
- Identify of SAR SDUs
- Provide error indication and handling
- Ensure SAR SDU sequence continuity
- Provide multiplexing and demultiplexing

The cell structure of the ATM AAL3/4 packet (Table 14.3) includes:

- Common Part Convergence Sublayer (CPCS)
  - Header
    - Message Type (CPI)
      - Set to zero when the BASize and Length fields are encoded in bytes
    - Beginning Tag (Btag)
      - Packet Identifier
      - Repeated in Etag
    - Buffer Allocation Size (BASize)
      - Receiver-allocated byte sizes to capture all the data
    - CPCS SDU
      - Variable information field
      - Varies from 0 to 65535 bytes
  - Information
  - Trailer
    - Padding (PAD)
      - 32-bit alignment of the length of the packet
  - All-zero

- End tag (Etag)

    - Equal to Btag

- Length

    - Must equal BASize

- Service Specific Convergence Sublayer (SSCS)

    — Header

    - Segment Type (ST)

        - Beginning of message (BOM) = 10

        - Continuation of message (COM) = 00

        - End of message (EOM) = 01

        - Single segment message (SSM) = 11

    - Sequence Number (SN)

        - SAR PDUs of a CPCS PDU number stream

    - Multiplexing Identification (MID)

        - Multiplexes several AAL3/4 connections over one ATM link

    — Information

    - Fixed length of 44 bytes

    - Contains parts of CPCS PDU

    — Trailer

    - Length Indication (LI)

        - SAR SDU length in bytes

        - BOM, COM = 44

        - EOM = 4, ..., 44

        - EOM (Abort) = 63

        - SSM = 9, ..., 44

    - Cyclic Redundancy Check (CRC)

TABLE 14.3 Cell Structure of AAL 3/4

Common Part Convergence Sublayer (CPCS)								Service Specific Convergence Sublayer (SSCS)					
Header	Information				Trailer			Header			Infor-mation	Trailer	
CPI	Beg. Tag (Btag)	Buffer Alloc. Size (Basize)	CPCS SDU	Pad	0	End Tag (Etag)	Length	Seg. Type (ST)	Seq. No. (SN)	Mux ID (MID)		Length Indica-tion (LI)	Cyclic Redun-dancy Check (CRC)
1	1	2	0— 65535	0—3	1	1	2 bytes	2	4	10	352	6	10

AAL5

The Adaptation Layer (AAL5) is a simplified version of AAL3/4. It is the most popular AAL used and has gained the name *Simple and Easy Adaptation Layer* (SEAL). This service consists of:

- Message and streaming modes
- ATM point-to-point and point-to multipoint connections
- Computer data transmission (i.e., TCP/IP)

The cell structure of the ATM AAL5 CPCS PDU (Table 14.4) contains:

- CPCS Payload
  - Actual information sent by the user
  - Information comes before any length indication
    - As opposed to AAL3/4, amount of memory required is known in advance
- Padding

— Adjusts the entire packet to fit into a 48-byte boundary

- CPCS User-to-User Indication (UU)

  — Transfers one byte of user information

- Common Part Indicator (CPI)

  — Reserved for future use

  — Value = 0

  — Will be used for layer management message indication

- Length

  — User information length without the pad

- Cyclic Redundancy Check (CRC)

  — Allows identification of corrupted transmission

TABLE 14.4

AALS Cell Structure

Information	Trailer				
CPCS Payload	Padding	User-to-User (UU)	Common Part Indicator (CPI)	Length	Cyclic Redundancy Check (CRC)
0—65535	0—47	1	1	2	4 bytes
Information	**Trailer**				
Information	Padding	User-to-User (UU)	Common Part Indicator (CPI)	Length	Cyclic Redundancy Check (CRC)
0—48	0—47	1	1	2	4 bytes

ATM Network Topology

A network switch is the root of ATM connectivity. This switch dictates the path of call from source to destination. Negotiation between the switch and node takes place to determine an open path to the destination node. The sending node indicates the type of data to be sent, required transmission speed, and other information about the requested transmission. This information determines the type of transmission channel to be made available to the node (e.g., higher speed, additional bandwidth, etc.). ATM topology is illustrated in Figure 14.3.

Figure 14.3
ATM topology.

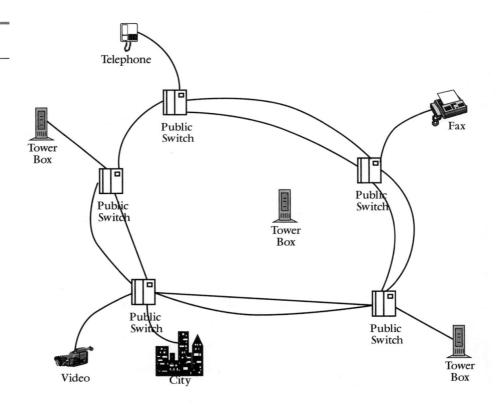

Advantages of ATM Network Topology

The advantages of ATM switching, when compared to network topologies consisting of shared technologies include:

- Use of higher bandwidths

- Data transmission at access speeds appropriate to the type of data sent

- Each telecommunication session contains its own dedicated bandwidth

- Support for point-to-point connection processes

ATM Connectivity

ATM connectivity (shown in Figure 14.4) through a network switch requires certain procedures to dictate the path that a cell can take from source to destination:

- Node negotiates with the switch for an open path to the destination node

- The sending node indicates:
  - Type of data to be sent
  - Transmission speed needed

- Sending node information determines the type of transmission channel to be made available to the node

- Switching technology
  - Transmits various types of data transfer needs
  - Enables data to be transmitted at access speeds appropriate to the type of data sent
  - Allows higher bandwidths
  - Enables each ATM communication session to have its own dedicated bandwidth

— Has clearly defined connection processes

— Handles point-to-point transfer

Figure 14.4
ATM access method
and connectivity.

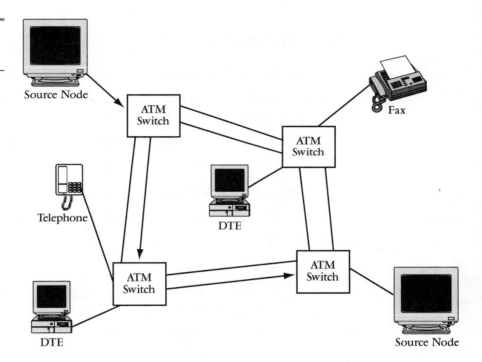

ATM Physical Interfaces

ATM offers various physical interfaces to fit specific requirements for speed, bandwidth, and other parameters. The most common interfaces used today are:

- DS-1
- DS-3
- E1
- E3

- Sonet OC-3c/SDH STM-1

- Taxi

- 25 Mbps

DS-1 Interface

- Operates at 1.544 Mbps over UTP-3 cables

- Compliant with ATM Forum User-to-Network Interface (UNI) specifications

- Supports direct cell mapping or cell delineation

  — Process of framing to ATM cell boundaries

- Supports Physical Layer Convergence Protocol (PLCP)

  — Provides the transmission of 10 ATM cells every 3 msec

- DS-1 frame is 193 bits long

  — First bit is overhead

  — Remaining 192 bits are made up of 8 bits of payload from each of 24 users ($8 \times 24 = 192$ bits)

- 12 frames are transmitted together as a Superframe (SF)

- 24 frames may be transmitted together as an Extended Super-frame (ESF)

TABLE 14.5   DS-1 Interface Frame Structure

F-bit	Payload							
First bit (F)	**User 1**	**User 2**	**User 3**	**User 4**	**User 5**	**User 6**		**User 24**
1 bit	8 bits	8 bits	8 bits	8 bits	8 bits	8 bits		8 bits
1 bit	192 bits							

DS-3

- Operates at 44.736 Mbps over coax cables
- Compliant with ATM Forum UNI specifications
- Supports direct cell mapping or cell delineation
  - Process of framing to ATM cell boundaries
- Supports PLCP
  - Provides the transmission of 12 ATM cells every 125 usec, or 4,608 Mbytes/sec (net transmission)
- Supports C-bit framing
  - Far-end performance monitoring
  - Far-End Alarm and Control signal (FEAC)
  - DS-3 path parity information
  - Far-end block errors
  - Path maintenance data link (using LAPD) on DTE to DTE
- Interface to BNC connectors
- Three framing standards
  - C-bit parity
    - 1 block
    - Data not muxed
    - Uses C-bits for purposes other than bit stuffing
  - M23 multiplex scheme
    - Provides for transmission of 7 DS-2 channels or 28 DS-1 channels
      - Each DS-2 channels contains 4 DS-1 channels
  - SYNTRAN
    - Multistep
    - Partially synchronous
    - Partially asynchronous multiplexing sequence

- Structure consists of:
  - M-frame partitions of 4,760 bits each
  - M-frames divided into seven M-subframes of 680 bits
  - Subframes divided into eight blocks of 85 bits each
    - First block is used for control, the rest is payload
  - 56 overhead bits that handle:
    - M-frame alignment
    - M-subframe alignment
    - Performance monitoring
    - Alarm channels
    - Application channels

TABLE 14.6

DS-3 Interface Frame Structure

X	PL	F (1)	PL	CB	PL	F (0)	PL	CB	PL	F (0)	PL	CB	PL	F (1)	PL
X	PL	F (1)	PL	CB	PL	F (0)	PL	CB	PL	F (0)	PL	CB	PL	F (1)	PL
P	PL	F (1)	PL	CB	PL	F (0)	PL	CB	PL	F (0)	PL	CB	PL	F (1)	PL
P	PL	F (1)	PL	CB	PL	F (0)	PL	CB	PL	F (0)	PL	CB	PL	F (1)	PL
M (0)	PL	F (1)	PL	CB	PL	F (0)	PL	CB	PL	F (0)	PL	CB	PL	F (1)	PL
M (0)	PL	F (1)	PL	CB	PL	F (0)	PL	CB	PL	F (0)	PL	CB	PL	F (1)	PL
M (1)	PL	F (1)	PL	CB	PL	F (0)	PL	CB	PL	F (0)	PL	CB	PL	F (1)	PL
M (0)	PL	F (1)	PL	CB	PL	F (0)	PL	CB	PL	F (0)	PL	CB	PL	F (1)	PL

PL = payload; Payload = 84 bits

E1 Interface

- Operates at 2 Mbps over coax cables

- Compliant with ATM Forum UNI specifications

- Supports direct cell mapping

  — ATM cells are carried in bits 9—28 and 137—256, which correspond to channels 1—15 and 17—31

- Supports PLCP

  — 10 rows of 57 bytes each

  — For overhead purposes, 4 bytes are added to the cell length of 53 bytes

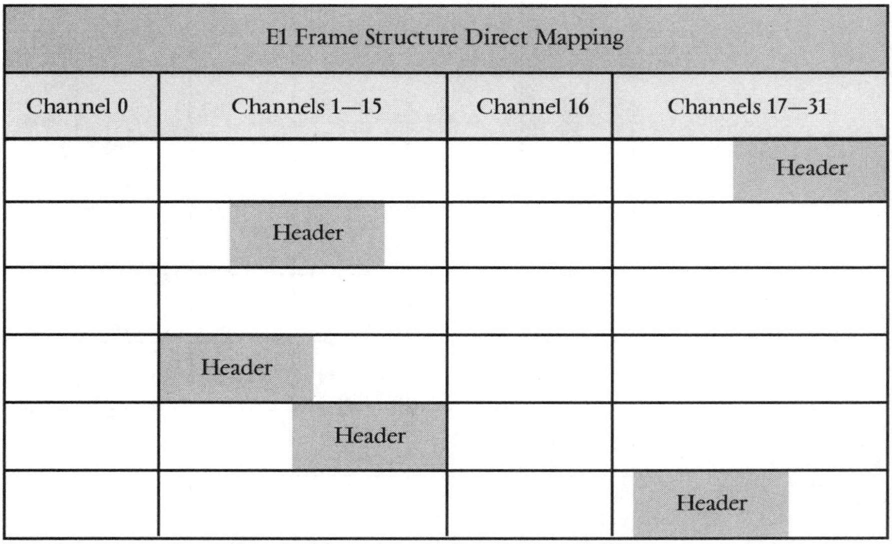

Figure 14.5
E1 interface frame structure.

E1 Frame Structure Direct Mapping			
Channel 0	Channels 1—15	Channel 16	Channels 17—31
			Header
	Header		
	Header		
	Header		
			Header

E3 Interface

- Operates at 34.368 Mbps over coax cables

- Compliant with ATM Forum UNI specifications

- Supports direct cell mapping

  — No relationship between the start of a direct mapping frame and the start of the ATM cell

- Supports PLCP

  — Nine ATM cells every 125 usec

  — Net transmission rate is 3.456 Mbytes/sec

  — For overhead purposes, 16 bytes are added to the cell length of 53 bytes

Figure 14.6
E3 interface frame structure.

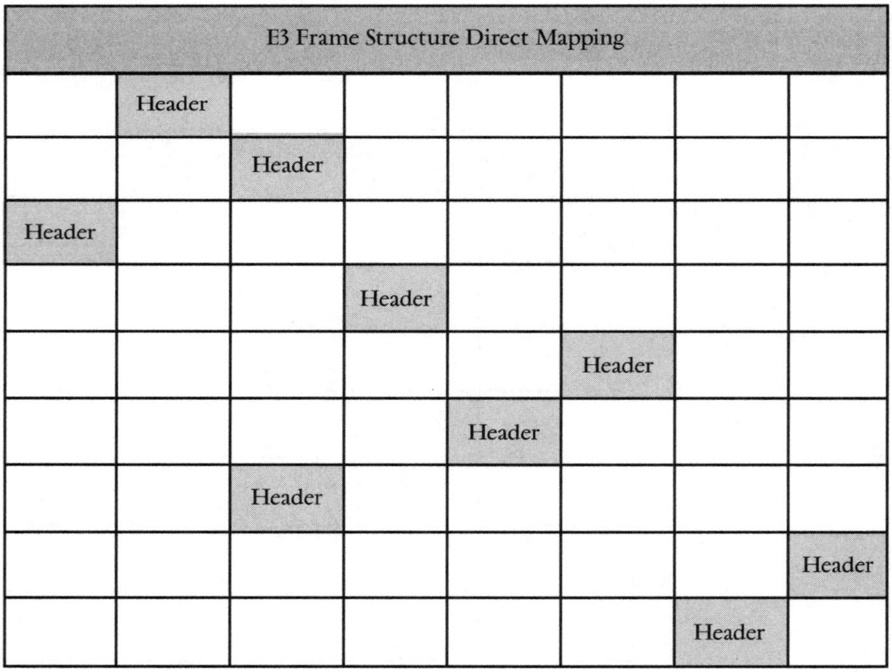

E3 Frame Structure Direct Mapping							
	Header						
		Header					
Header							
			Header				
					Header		
				Header			
		Header					
							Header
						Header	

Synchronous Optical Network (SONET) OC-3c/Synchronous Digital Hierarch (SDH) STM-1 Interface

- Operates at 155 Mbps over SONET or SDH interfaces
- SONET is the most widely used interface with ATM
- Compliant with ATM Forum UNI 3.0 specifications
- Connections may be:
  - Multimode
    - Uses SC-type optical connectors
  - Single-mode
    - Uses SC-type optical connectors
  - UTP
    - Uses UTP-5 connectors
- Both SONET and SDH are based on transmission at speeds of multiples of 51.840 Mbps or STS-1
- OC-3c and STM-1 rates are an extension of the basic STS-1 speed, which operates at 155.520 Mbps
- Payload may float inside the OC-3c frame in case the clock used to generate the payload is not synchronized with the clock used to generate the overhead
- Actual useful information rate carried inside the OC-3c payload is 149.76 Mbps
  - 5 bytes out of every 53-byte cell are the header
  - Only 135.63 Mbps carry actual ATM payload

25 Mbps Interface

- Operates at 25.6 Mbps over twisted pair cables
- Compliant with ATM Forum UNI specifications

- Interface has 4B/5B Nonreturn to Zero (NRZI) coding
- Uses RJ48 electrical connectors

TAXI Interface

- Operates at 100 Mbps over multimode fiber transmission
- Specified in the ATM Forum, ATM User—Network Interface Specifications 3.0
- Takes advantage of the FDDI LAN systems:
  - Uses existing chips of cell transport
  - Uses same physical media
  - Uses same lasers
  - Uses AMD TAXI chips
  - Does not use ring architecture
- Links
  - Full-duplex
  - Point-to-point
  - Carrying 53-byte ATM cells with no physical framing structure

ATM Signaling

ATM signaling exchanges data between the users and the network. This exchange data includes:

- Control information
- Network resource requests
- Circuit parameter negotiation

Required bandwidth is allocated as a result of a successful signaling exchange.

The ATM signaling protocols are transmitted over the Signaling ATM Adaptation Layer (SAAL). The SAAL:

■ Ensures reliable delivery

■ Is divided into four parts:

— Service Specific Part

— Service Specific Coordination Function (SSCF), which interfaces with SSCF user

— Service Specific Connection-Oriented Protocol (SSCOP), which assures reliable delivery

— Common Part

Figure 14.7 illustrates the relationship between the ATM Signaling Protocol Stack and the Signaling ATM Adaptation Layer (SAAL).

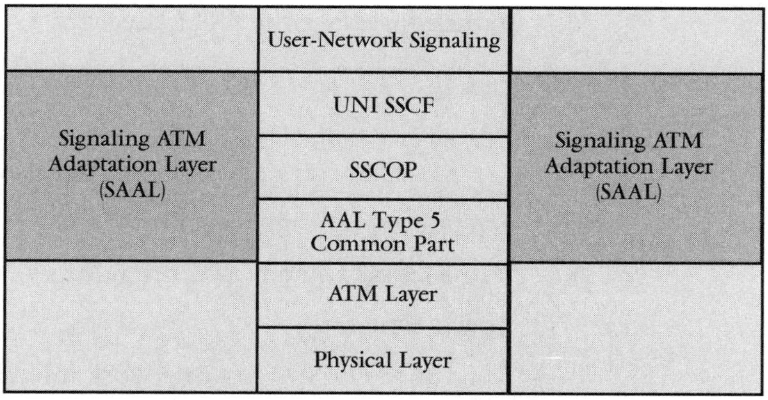

Figure 14.7
Relationship of ATM signaling protocols to the Signaling ATM Adaptation Layer (SAAL).

Signaling ATM Adaptation Layer (SAAL)	User-Network Signaling	Signaling ATM Adaptation Layer (SAAL)
	UNI SSCF	
	SSCOP	
	AAL Type 5 Common Part	
	ATM Layer	
	Physical Layer	

The protocols within the SAAL are responsible for:

■ ATM calls

■ Connection control

■ Call establishment

■ Call clearing

- Status inquiry

- Point-to-multipoint control

ATM Protocol Variations

Several variations of ATM signaling protocols exist. Each protocol contains its own characteristics and features. Based on which protocol is chosen, data fields will vary. It would be a book in itself to detail all the information regarding these protocols, so we just list the more popular protocols and their descriptions:

- BISDN InterCarrier Interface (B-ICI)

  — Interface connecting two different ATM-based public network providers or carriers

  — Facilitates end-to-end national and international ATM/BISDN services

  — Functions above the ATM Layer to:

    - Transport

    - Operate

    - Manage a variety of intercarrier services across the B-ICI

- Interim Interswitch Signaling Protocol (IISP)

  — Provides signaling between vendor switches

- ITU Q.2931 Signaling

  — Used at the B-ISDN user-network interface

  — Specifies procedures for establishment, maintenance, and clearing of network connections

  — Procedures are defined in terms of messages exchanged

- Multiprotocol over ATM (MPOA)

  — Without requiring routers, allows intersubnet to internetwork layer protocol

    - Transfers intersubnet unicast data

    - Preserves benefits of LAN emulation

- — Provides framework for:
    - Diverse protocols
    - Network technologies
    - IEEE 802.1 virtual LANs
- — Communicates with routers and bridges to determine optimal exit from the ATM cloud

- Private Network-to-Network Interface (PNNI)
    - — Hierarchical, dynamic link-state routing protocol
    - — Supports large-scale ATM networks
    - — Supports connection establishment across multiple networks
    - — Dynamically establishes, maintains, and clear ATM connections at the following locations:
        - Private network-to-network interface
        - Network-node interface between two ATM networks
        - Two ATM network nodes

- Simple Protocol for ATM Network Signaling (SPANS)
    - — Developed by FORE Systems; used on FORE Systems and other compatible networks
    - — Uses AAL3/4 to transfer over a reserved ATM virtual connection
    - — Retransmission of lost messages and suppression of duplicate messages is performed by the application
    - — Null transport layer is used

- UNI 4.0 Signaling
    - — At the ATM user—network interface, provides signaling procedures for dynamically establishing, maintaining, and clearing ATM connections
    - — Public interfaces between endpoint equipment and a public network
    - — Private interfaces between endpoint equipment and a private network

- ViVID Multiprotocol over ATM (MPOA)

  — Proprietary protocol of Newbridge Corporation

  — Provides bridged LAN Emulation (LANE) functionality

  — Provides routed LAN Emulation (LANE) functionality

ATM Encapsulation Procedures

ATM offers the capability to integrate ATM into existing LANs and WANs. ATM integrates a number of standards that describe the encapsulation of LAN and WAN protocols over ATM. Two of the most common protocol standards used over ATM are:

- Frame Relay over ATM
- IP Addressing over ATM

There are two methods used to encapsulate or transport LAN and WAN protocols via ATM:

- Virtual Channel-Based Multiplexing

  — Uses one virtual channel for each protocol

  — Transmitted over the AAL5 PDU

  — No additional payload required

  — Used on routed protocols such as:

    - TCP/IP, where the PDU is carried directly in the payload of the AAL5 CPCS PDU

  — Used on bridged protocols such as:

    - Token Ring and Ethernet, where the PID field is carried in the payload of the AAL5 CPCS PDU

- Multiprotocol Encapsulation over ATM Adaptation Layer 5

  — Encapsulation of LAN protocols over ATM with use of header value

LAN or WAN emulation protocol uses control messages to set up the LAN. LAN Emulation (LANE) supports two possible data packet formats:

- Ethernet

- Token Ring

LAN emulation data frames preserve all the information contained in the original 802.3 or 802.5 frames, but adds a 2-byte Local Exchange Carrier (LEC) source ID, which is unique to each Local Exchange Carrier (LEC).

ATM Circuit Emulation

ATM consists of many advantageous features (such as increased speed, bandwidth, etc.) that encourage the development of standard protocols for transmittal of video and audio over ATM. The following standards are popular choices for transferring these signals over ATM:

- ATM Circuit Emulation

  - Provides connection between Constant Bit Rate (CBR) equipment across an ATM network in a transparency mode

- Digital Storage Media Command and Control (DSM-CC)

  - Provides the control functions and operation specific to managing bit streams

- MPEG-2

  - Compressed representation of video and audio sequences using a common coding syntax

  - Ability to support voice, video, and data simultaneously

  - Specifies the coded bit stream for high-quality digital video

  - Supports interlaced video formats for:

    - Cable television (CHTV)

    - Broadcast satellite (DBS)

    - High Definition TV (HDTV)

Digital Subscriber Line (DSL)

Digital Subscriber Line (DSL) technology provides high-speed, non-switched digital data transport. DSL is designed to connect an end user to an Internet Service Provider (ISP) or a corporate network intranet.

To get down to basics, DSL signifies a *modem pair*. DSL requires one modem and a line requires two modems—one at each end of the communications line. A modem pair and appropriate software applied to a line creates a digital subscriber line.

A DSL modem is similar to the modem used for Basic Rate ISDN—in fact, some may argue that it is the same modem. DSL transmits data in both directions simultaneously at 160 Kbps over copper lines. The distance on these copper lines may run to 18,000 feet on 24-gauge wire. Data streams are multiplexed and demultiplexed into two B channels (64 kbps each), a D channel, and some overhead. DSL is a standard implementation (ANSI T1.601 or ITU I.431) that employs echo cancellation to separate the transmitted signal from the received signal at both ends. Figure 15.1 illustrates a DSL-to-ISDN connection.

Figure 15.1
DSL-to-ISDN
connection.

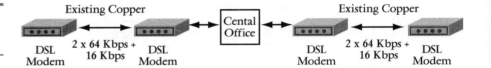

DSL modems use twisted-pair cable supporting bandwidth from 0 to 80 kHz. DSL modems preclude the simultaneous provisioning of analog Plain Old Telephone Service (POTS). Today, DSL is used for pair-gain applications because DSL modems convert a single POTS line into two POTS lines thus removing the necessity of installing a second wire.

DSL Advantages

The advantages of DSL deployment include:

- High-speed transport for small business customers, corporate telecommuters, or high-end residential customers

- Lower speed 128 Kbps to 388 Kbps and 1.5 Mbps with a minimum downstream speed of 388 Kbps
- Provides high-quality video files, animation features, color-rich graphics, and large data-file transfer from work to home or vice-versa over a voice line
- Ten to fifty times faster than ISDN or analog dial-up
- Efficient use of cost and time
- Provides office presence from a remote location
- Provides required bandwidth for applications
- Provides immediate access to Internet
- Future offerings may include:
  - Security functionality and features
  - Switched virtual circuits
  - Protocol conversion
- Future transport applications may include:
  - IP-based virtual private networks (VPNs)
  - Multiple digital voice lines in DSL spectrum
  - Remote medical imaging
  - Broadband electronic commerce
  - Videoconferencing and videoconferencing bridges

DSL Service Features

- Support for ISP applications
- Customer's DSL modem contains a rate-adaptation feature to coordinate a "handshake" with the Central Office equipment
- High level of security
- Service with dedicated Central Office access
- ATM backbone
- Customers have choice of ISPs

DSL Requirements

DSL Network Requirements

An itemized list of criteria must be met before deploying DSL to the customer:

- The telecommunication Central Office in which the customer's residence or business location is served must be deployed with DSL equipment

- The customer's location must be within a specific loop-length from the central office; today, the loop-length is 12,000 feet from the DSL Central Office

- The line may require line conditioning, which may include the removal of

  - Load coils

  - Bridge taps

  - Repeaters

- DSL service is utilized on top of the ATM platform

  - ATM is required to be provided by the Central Office and Internet Service Providers (ISPs)

- Internet Service Providers (ISPs) and corporate LAN network customers require the following:

  - High-speed connection over a ATM cell relay service

  - Connectivity ranging from DS-1 to OC-3 levels

  - Connection need not be dedicated to DSL

  - Hosts sites require bandwidth

  - Hosts sites require the appropriate CPE to support high-speed transport

Customer Equipment Requirements

Along with network requirements, it is necessary that the customer's equipment also meet certain requirements:

- Customers must have a personal computer (PC):
    - Pentium Processor (I, II, III, IV, etc.)
    - Windows 95 operating system or greater
    - 16 Mb of Random Access Memory (RAM)—however, 32 Mb is recommended
    - Vacant slot for NIC card
    - Original CD-ROM or diskettes for the operating system
    - Internet software
    - DSL modem
    - 25 Mb disk space on hard drive
- If a Macintosh computer is used:
    - Hardware—68030 or greater
    - Software—7.0 or better
    - Vacant slot for NIC card
    - Original CD-ROM or diskettes for the operating system
    - Internet software
    - DSL modem
    - 25 Mb disk space on hard drive

Service Components

Telecommunication service components necessary for DSL deployment include:

- Plain Old Telephone Service (POTS)
    - Customer must have a basic telephone line

- Rate element
  - USOC must provision the actual speed (384/128 or 384/1.5M)
- POTS splitter
  - Allows POTS and DSL signals to coexist on the same twisted-pair
- DSL modem
  - Present at customer's end of the line
  - Connects PC to the customer's ISP or remote LAN
  - Provides proper dialing standards
- Network Interface Card (NIC)
  - Either an ATM/Ethernet PC card or an Ethernet/Macintosh card
- ATM cell relay service
  - Internet Service Provider (ISP) or corporate LAN must have a connection to an ATM edge switch via an ATM cell relay service facility

Data Speed Factors

Several factors can affect data speeds at a particular service address:

- Media used for transport
  - ISDN
  - T1
  - T3
  - Etc.
- Load coils, SLC, bridge taps, etc.
- Length and gauge of copper loop
- Impulse noise or environmental interference
- Wiring on customer premises

- DSL has no impact on an existing POTS service

  — DSL and POTS services use different frequencies.

  — Data and voice may be transported at the same time

DSL Application Support

DSL supports two primary applications on the end user side

- Remote office connection
- Internet access

Using DSL, the end user has a high-speed access to an ISP or a corporate LAN network. Table 15.1 shows the approximate time required to download a 10 Mb file using various transport architectures.

TABLE 15.1

Download Time
Comparison for
DSL

Analog Modem (28.8 Kbps)	5 minutes 47 seconds
ISDN (128 Kbps)	1 minute 18 seconds
DSL (384 Kbps)	26 seconds
DSL (1.5 Mbps)	7 seconds

DSL Procedures

DSL Connection Procedures

Customers must establish a connection by following these steps:

- Dedicated connection from the home to the Central Office
- At the Central Office, the data traffic is multiplexed
- After multiplexing, traffic is routed to the data network

- Customer must have a dedicated connection through a Virtual Communications Circuit (VCC) to:

  — An Internet Service Provider (ISP)

  — An Online Service Provider (OSP)

  — A corporate enterprise network with a *Virtual Communications Circuit* (VCC)

- VCC is connected from the Digital Subscriber Line Access Multiplexer (DSLAM) to the ATM network

- Customer finally connects to an ISP or corporate network

DSL Transmission Procedures

- Offer the capability of transmitting data over voice services simultaneously using the same POTS network on twisted-pair cable

- Once data reaches its destination (the end user or Central Office location) a splitter is required to separate the data traffic from the voice traffic

- Voice traffic is routed to the voice switch and the data traffic goes through the Digital Subscriber Line Access Multiplexer (DSLAM) for multiplexing and connection to the fast packet network

- The end-user will receives the higher speeds in the downstream direction, which provides ideal service for Internet and host-remote applications

- DSL bypasses the voice switch for data sessions and places data sessions on a data network that is better equipped to handle continuous and more demanding data flows

Figure 15.2 illustrates a typical DSL network architecture.

Figure 15.2
DSL architecture.

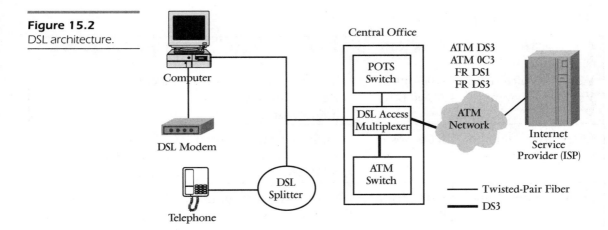

DSL Protocols

There are variations in the DSL protocol depending on data rate, applications, and other variable factors. Table 15.2 describes the most commonly used DSL services utilized today.

Bandwidth Limitations

Bandwidth limitations occur in the core network. Voice grade bandwidth is limited to 3.3 kHz by filters located at the edge of the core network. With attenuation, copper access lines can pass frequencies into MHz regions. Attenuation increases with line length and frequency and dominates the constraints on data rate over twisted-pair wire. Table 15.3 maps data rate limits to line length.

To increase the length line distance requirement for DSL, telephone companies are working to shrink the loop length. A technique to stretch the capacity of the existing central office involves installation of access nodes located remote from the central office.

- Remote sites are called *distribution areas*

- They carry a maximum subscriber loop of 6,000 feet from the access node

- They are fed by T1 or E1 lines or fiber, using the HDSL protocol

TABLE 15.2   DSL Protocol Variations

Name	Meaning	Data Rate	Mode	Applications
V.22 V.32 V.34	Voice band modems	1200 bps to 28,800 bps	Duplex—Data at same rate both upstream and downstream	Data communications
DSL	Digital Subscriber Line	160 Kbps	Duplex—Data at same rate both upstream and downstream	ISDN service Voice and data communications
HDSL	High Data Digital Subscriber Line	1.544 Mbps—Requires two twisted-pair lines 2.048 Mbps—Requires three twisted-pair lines	Duplex—Data at same rate both upstream and downstream	T1/E1 service Feeder plant, WAN, LAN access, server access
SDSL	Single Line Digital Subscriber Line	1.544 Mbps 2.048 Mbps	Duplex—Data at same rate both upstream and downstream	T1/E1 service Feeder plant, WAN, LAN access, server access, premises access for symmetric services
ADSL	Asymmetric Digital Subscriber	1.5 to 9 Mbps 16 to 640 Kbps	Downstream—Network to subscriber Upstream—Subscriber to network	Internet access, video demand, simplex video, LAN access, interactive multimedia
VDSL	Very High Data Rate Digital Subscriber Line, also referred to as BDSL, VADSL, or ADSL. (VDSL is ANSI and EISI designation.)	13 to 52 Mbps 1.5 to 2.3 Mbps	Downstream—Network to subscriber Upstream—Subscriber to network	Internet access, video demand, simplex video, LAN access, interactive multimedia, HDTV

TABLE 15.3

Data Rate-to-Line
Length
Comparison

Wires	Data Rate Limits	Line Length
DS1 (T1)	1.544 Mbps	18,000 feet
E1	2.048 Mbps	16,000 feet
DS2	6.312 Mbps	12,000 feet
E2	8.448 Mbps	9,000 feet
¼ STS-1	12.960 Mbps	4,500 feet
½ STS-1	25.920 Mbps	3,000 feet
STS-1	51.840 Mbps	1,000 feet

T1 and E1 Circuits

Originally, telephone companies used T1 and E1 circuits for transmission between offices in the core switching network. Today, T1 and E1 have become tariffed, so the services currently being offered are:

- Private networks implementation
- Connection of T1 multiplexers over the Wide Area Network (WAN)
- Connection of Internet routers
- Establishment of connections for traffic from a cellular antenna Central Office
- Connection of multimedia servers into a Central Office
- Feeder plant implementation
  - Feeds digital loop carrier systems
  - Concentrates 24 or 30 voice lines over two twisted-pair lines from a Central Office
  - Reduces distance between an access point and the final subscriber

T1 and E1 Limitations

T1 and E1 service does not work well for connecting to individual residences:

- *Alternate Mark Inversion* (AMI) protocol, which was originally used on a T1/E1 service

  — Demands a lot of bandwidth

  — Corrupts cable spectrum

  — Requires the replacement of all lines to a customer's site

- Very few applications going to a home demands such a high data rate

- Increasing data rate requirements are accelerating the demands as highly asymmetric with very little upstream in return and require rates above T1 or E1

- High speed services to the home are carried by ADSL or VDSL

High Data Rate Digital Subscriber Line (HDSL) Protocol

High Data Rate Digital Subscriber Line (HDSL) also transmits T1 or E1 over twisted-pair copper lines. HDSL is:

- More mature than other DSL technologies

- Utilizes data rates above 1 megabit

- May be utilized for premises applications for Internet remote LAN access

- Uses less bandwidth

- Requires no repeaters

- Uses advanced modulation techniques

- Supports typical applications such as:

— Cellular antenna stations

— PBX network connections

— Digital loop carrier system interexchange POPs

— Private data networks

— Internet servers

If higher data rates are required, ADSL and SDSL are the next steps up in the DSL hierarchy.

Symmetric Digital Subscriber Line (SDSL) Protocol

Symmetric Digital Subscriber Line (SDSL) is a single line version of HDSL. SDSL:

- Transmits T1 or E1 signals over a single pair
- Operates over POTS
- Supports single-line POTS and T1 and E1 simultaneously
- Works well for individual subscriber premises because it only requires one line
- Supports applications requiring symmetric access
- Complements ADSL

Asymmetric Digital Subscriber Line (ADSL) Protocol

Asymmetric Digital Subscriber Line (ADSL) (see Table 15.4) transmits an asymmetric data stream. More transmission goes downstream (possibly 1.5 or 3.0 Mbps) to the subscriber and much less goes upstream (possibly 64 Kbps). Upstream data rates usually range from **16 to 640** Kbps. The targeted applications for this service are:

- Multiple switching networks:
    - Circuit switched
    - Packet switched
    - ATM switched
- Internet access
- Transmission of digitally compressed video
- Error correction capabilities
- Reduction in effect of impulse noise on video signals
- Remote LAN access
- Ability to connect multiple applications simultaneously:
    - Multimedia access
    - Specialized PC service
    - Video demand

TABLE 15.4

ADSL Downstream
Data Rates

Line Service	Data Rate	Line Distance
DS1 (T1)	1.544 Mbps	18,000 feet
E1	2.048 Mbps	16,000 feet
DS2	6.312 Mbps	12,000 feet
E2	8.448 Mbps	9,000 feet

Very High Data Rate Digital Subscriber Line (VDSL) Protocol

Very High Data Rate Digital Subscriber Line (VDSL) uses asymmetrical transceivers at data rates higher than ADSL (see Table 15.5). Upstream rates are within the range of 1.6 Mbps to 2.3 Mbps.

TABLE 15.5

VDSL Downstream
Data Rates

Line Service	Data Rate	Line Distance
$\frac{1}{4}$ STS-1	12.960 Mbps	4,500 feet
$\frac{1}{2}$ STS-1	25.920 Mbps	3,000 feet
STS-1	51.840 Mbps	1,000 feet

Compared to ADSL, VDSL is:

- Simpler than ADSL

- Has fewer transmission constraints than ADSL because of its shorter lines

- Uses transceiver technology less complex than ADSL

- Implements a data rate 10 times faster than ADSL

The characteristics of VDSL are:

- Operates on ATM, POTS, and ISDN network architectures

- Admits passive network terminations

- Enables more than one VDSL modem to be connected to the same line at a customer premise

- Provides error correction

- Support switched networks:

  — Circuit switched

  — Packet switched

- VDSL is also referred to as VADSL, BDSL, or ADSL—old terms for this service that may still appear in references

Synchronous
Optical Network
(SONET)

Synchronous Optical NETwork (SONET) is a standard for optical telecommunication transport. SONET Ring and Access Service is available out of both the CPUC and FCC tariffs. This service has the highest level of redundancy and network availability.

Before SONET, fiber-optic systems in the public telephone network used proprietary architectures, equipment, line codes, multiplexing formats, and maintenance procedures. User demanded standards so that they could mix and match equipment from different vendors. The task of developing such a standard was handled by the ECSA organization for ANSI in 1984. Their mission was to establish a standard for connecting one fiber system to another.

ECSA sets industry standards in the United States for telecommunications and other industries. The SONET/SDH (Synchronous Digital Hierarchy) standard is expected to provide the transport infrastructure for worldwide telecommunications for at least the next two or three decades. Also, SONET utilizes an internationally recognized (ITU/CCITT) framing standard.

SONET defines Optical Carrier (OC) levels and Synchronous Transport Signals (STSs) for the fiber optic-based transmission hierarchy.

SONET Advantages

Implementation of SONET provides significant advantages over older telecommunications systems, including:

- Reduction in equipment requirements
- Scalable access service
  - Declining cost per data byte with greater bandwidth options
  - Flat-rate customer pricing—charges are not usage sensitive
  - Guaranteed lowest rates through RSPP
- Flexibility to grow or upgrade
  - Additional bandwidth or roll over to other access services does not incur termination liability charges
  - Architecture is capable of accommodating future applications

- — Accommodates a variety of transmission rates
- Increased network reliability
- Centralized fault monitoring
  - — Permits management of payload bytes on an individual basis
- Synchronous multiplexing format capability
  - — Can carry lower level digital signals
  - — Simplifies the interface to:
    - Digital switches
    - Add-drop multiplexers (ADMs)
    - Digital cross-connect switches
- Economies of scale
  - — Used to economically aggregate subrate access services such as ADNs/DS0s, T1s/DS1s, DS3, OC3c, and OC12c onto one transport network platform
  - — A single network is less expensive and more efficient to operate than multiple networks.

SONET Hardware and Software Integration Advantages

SONET supports the integration of multiple applications onto an integrated access backbone that the customer's entire access network can utilize:

- Integrated platform provides access to local voice, data, and video services through telecommunication Central Offices, frame, and cell relay switches
- Integrated multiple applications that provide Internet and private network connectivity and access to long distance carriers are just a few of the applications that can ride an integrated access network

Fiber-to-Fiber Interfaces

SONET standards contain definitions for fiber-to-fiber interfaces at the physical level. Characteristics determined at this level are:

- Optical line rate
- Wavelength
- Power levels
- Pulse shapes
- Coding
- Frame structure
- Overhead
- Payload mappings

Standards on Integration

SONET standards integrate various vendors and applications using different optical formats. Various voice, video, and data applications are possible between networks with the use of SONET:

- High-speed internetworking (e.g., LAN, WAN interconnection)
- Large file transfer
- Aggregation of subrate circuits like AND/DS0, Digital Subscriber Line, T1/DS1, and OC3c providing connectivity to:
  - Internet Service Provider (ISPs)
  - Remote Local Area Network (RLAN) users
  - Distributed processing and file server access
- Multimedia
  - Full-motion video
  - Uncompressed video
- Videoconferencing

- — Access across Wide Area Networks (WANs) and Local Area Networks (LANs)
- Imaging
  - — Low-delay and high performance, which is necessary for image storage, retrieval, and transport
- Disaster recovery
  - — Remote access to sites
  - — Delivery of mirrored data
- High performance
  - — Provides a high degree of performance and reliability
  - — Service performance is backed by published network availability statistics and out-of-service credits
- Worldwide interoperability
  - — ATM service is based on international SONET standards, established by ANSI and International Telephone Union (ITU, formerly CCITT)

Multipoint Configurations

SONET supports a beneficial multipoint or hub configuration:

- Network providers are no longer required to own and maintain customer-located equipment
- Multipoint structure allows network providers and their customers to optimize their shared use of the SONET infrastructure
- Reduces the need for back-to-back terminals

Pointer, Multiplexer, and Demultiplexer

Network clocks are referenced to a highly stable reference point:

- Alignment of data streams is unnecessary

- Synchronization of clocks is unnecessary

- Lower rate signal is accessible

- Demultiplexing is not needed to access bit streams

- Signals can be stacked together without bit stuffing

- Flexible allocation and alignment of payload is permitted when frequencies vary with use of pointers

SONET Broadband Transport

Synchronous Optical Network (SONET) Ring and Access Service is a special access, private network service offering broadband transport at bandwidths above traditional asynchronous DS1 (1.544 Mbps) and DS3 (44.736 Mbps).

SONET speeds range from (STS1) 51 Mbps to OC48 (2.4 Gbps) using standard interfaces such as DS1, DS3, OC3c, and OC12c and provide a superior fiber-based architecture that virtually all other services and applications can ride.

SONET Ring and Access Service broadband transport comprises:

- Circuit Service

  — Point-to-point

  — BLSR SONET technology

  — DS1, DS3, or OC3c

- Dedicated Ring Service

  — Custom network built with full subscription to OC3, OC12, OC48, or OC192 bandwidths

  — Combination of DS1, DS3, OC3c, or OC12c circuits supported

Bidirectional Line-Switched Rings (BLSR) SONET Technology

Bidirectional Line-Switched Rings (BLSR) offer superior topology, redundancy, network availability, and failure response for mission-critical communications because of dual working and protection paths.

This extensive degree of availability is handled by having loopbacks at either side of the link failure, which reroutes traffic on and off the reserved protection bandwidth. The connection failure traffic is avoided by use of *K-byte* signaling in the SONET line overhead.

SONET contains a Centralized Network Management system. This network terminating equipment is classified as an intelligent network element. Network elements can be remotely monitored to send alarms as required.

SONET Dedicated Ring Service Components

Dedicated customer rings offer additional security, assured capacity, and personal customized design and planning to the customer's service requirements:

- Nodes

  — Provided by Central Office, Premise, and Customer

  — Operate as SONET add-drop multiplexer

  — Use dedicated facilities

  — Are limited to use by the recorded customer

- Links

  — Provide Interoffice Facilities, Local Loop, and Alternate Wire Center

  — Use two pairs of fiber optic transmission cable that connect the nodes on a ring

- Access Ports

  — Located at Central Office and Premise

  — Establish circuits on the ring through application handoffs

  — Support a variety of bandwidths: DS1, DS3, STS1, OC3, OC3c, OC12, and OC12c can be used to ingress and egress the ring

- Optional features
  - Term commitment
  - Bandwidth subscription
  - Central office multiplexing
  - Virtual tributary bandwidth management

SONET/SDH Technical Specifications

- High-speed optical transport service
- Can transmit subrate asynchronous circuits (e.g., DS1 and DS3)
- Operates at 155 Mbps over SONET or SDH interfaces
- Internationally recognized technology
- Internationally standardized rates with fixed-size signaling cells (1 cell = 53 bytes)
- Services supported
  - Asynchronous Transfer Mode (ATM) Cell Relay Service
    - Constant Bit Rate (CBR) applications for video and audio applications
    - Unspecified Bit Rate (UBR)
    - Variable Bit Rate (VBR) applications for data applications
  - Fast Packet Frame Relay Service applications
  - Switched services
    - Primary Rate ISDN (PRI-ISDN)
    - Super Trunk (digital entrance facilities)
- Most widely used interface with ATM
- Compliant with ATM Forum UNI 3.0 specifications
- Connections via

— Multimode, which uses SC-type optical connectors

— Single-mode, which uses SC-type optical connectors

— UTP, which uses UTP-5 connectors

- Both SONET and SDH are based on transmission at speeds of multiples of 51.840 Mbps or STS-1

- OC-3c and STM-1 rates are an extension of the basic STS-1 speed, which operates at 155.520 Mbps

- Payload may float inside the OC-3c frame in case the clock used to generate the payload is not synchronized with the clock used to generate the overhead

- Actual useful information rate carried inside the OC-3c payload is 149.76 Mbps

— 5 bytes out of every 53-byte cell are the header

— Only 135.63 Mbps carry actual ATM payload

SONET Signaling

SONET is a technology for transmitting various signals of different capacities through a synchronous, flexible, optical hierarchy. SONET accomplishes this task by utilizing a byte-interleaved multiplexing scheme that simplifies multiplexing and provides end-to-end network management. Table 16.1 shows the SONET signaling hierarchy. Table 16.2 illustrates a nonsynchronous hierarchy.

SONET multiplexing encompasses:

- The base signal or lowest level signal

— Synchronous Transport Signal-Level 1 (STS-1)

— Operates at 51.84 Mbps

- Higher level signals are integer multiples of STS-1

— Generates STS-N signals

- Composed of N byte-interleaved STS-1 signals

TABLE 16.1

SONET Hierarchy

Signal	Bit Rate	Capacity
STS-1, OC-1	51.840 Mbps	28 DS-1s or 1 DS-3
STS-3, OC-3	155.520 Mbps	84 DS-1s or 3 DS-3s
STS-12, OC-12	622.080 Mbps	336 DS-1s or 12 DS-3s
STS-48, OC-48	2,488.320 Mbps	1,344 DS-1s or 48 DS-3s
STS-192, OC-192	9,953.280 Mbps	5,376 DS-1s or 192 DS-3s

STS = Synchronous Transport Signal; OC = Optical Carrier

TABLE 16.2

Nonsynchronous Hierarchy

Signal	Bit Rate	Channels
DS-0	0.640 Mbps	1 DS-0
DS-1	1.544 Mbps	24 DS-0s
DS-2	6.312 Mbps	96 DS-0s
DS-3	44.736 Mbps	28 DS-1s

Transmission Control Protocol/ Internet Protocol (TCP/IP)

The Transmission Control Protocol/Internet Protocol (TCP/IP) was developed by the Department of Defense (DoD) under the aegis of the Advanced Research Project Agency Network (ARPANet). It has since been designated as the routing and end-to-end protocol supported not only on the DoD's Internet network, but on the Internet as a whole. Today we are looking for efficient ways to conjoin TCP/IP with traditional telephony to create third-generation telecom services.

Figure 17.1 compares the OSI Reference Model to the TCP/IP protocol and responsibility stack.

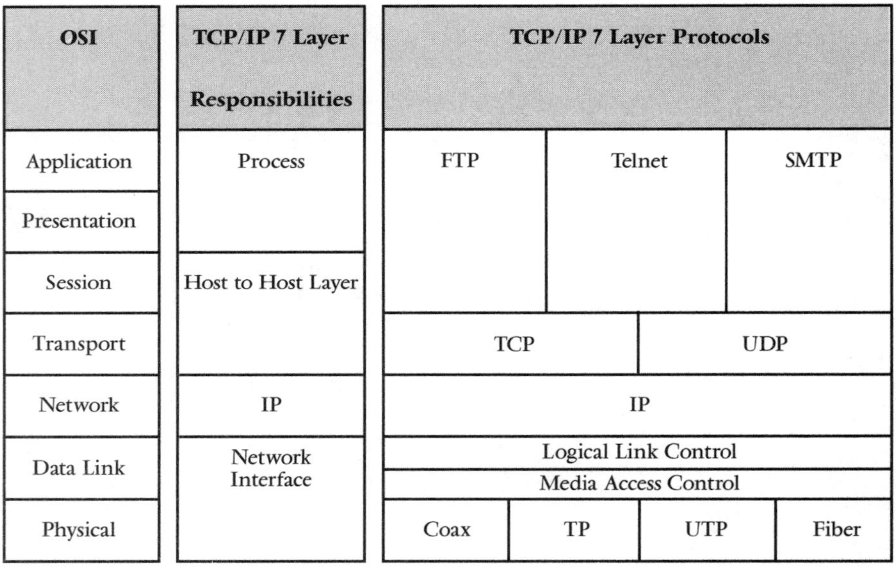

Figure 17.1
TCP/IP and the OSI Reference Model; a responsibility and protocol comparison.

OSI	TCP/IP 7 Layer Responsibilities	TCP/IP 7 Layer Protocols			
Application	Process	FTP	Telnet		SMTP
Presentation					
Session	Host to Host Layer				
Transport		TCP		UDP	
Network	IP	IP			
Data Link	Network Interface	Logical Link Control			
		Media Access Control			
Physical		Coax	TP	UTP	Fiber

Functionality of TCP and IP as Two Separate Entities

TCP/IP functionality is broken down into two areas for better definition:

- Transmission Control Protocol (TCP)
- Internet Protocol (IP)

Transmission Control Protocol (TCP)

Applications that require a transport protocol to provide reliable data delivery use TCP because it verifies that data is delivered across the network accurately and in the proper sequence. TCP is a *reliable, connection-oriented, byte-stream* protocol. TCP's strongest key point advantages are:

- Reliability
  - Supports error control procedures
  - Uses a mechanism called *Positive Acknowledgment with Retransmission* (PAR) to verify receipt of data
  - After a period of time, retransmits data unless the system receives positive acknowledgment from receiver that data has been received correctly
    - A *segment* is a unit of data exchanged between TCP modules
    - Each segment contains a *checksum* for verification that data is undamaged:
      - If the data segment is received undamaged, the receiver sends a positive acknowledgment back to the sender
      - If the data segment is damaged, the receiver discards it
- Connection-oriented format
  - Establishes a logical end-to-end connection between the two communicating hosts
  - Control information, called a *three-way handshake,* is exchanged between the two endpoints to establish a dialog before data is transmitted
    - Called a three-way handshake because three segments are exchanged
  - Transmitter sends a Synchronized (SYN) sequence number
  - Receiver responds with an Acknowledgment (ACK) segment and SYN bit set

— Transmitter acknowledges receipt of the receiver's segment and transfers the first actual data

— Connection is known to be established because both ends are aware that each is alive and ready for transmission of data

— When data transfers are concluded, sites exchange a three-way handshake with segments containing the "No more data from sender" bit (called the *FIN* bit) to close the connection

— End-to-end exchange of data provides the logical connection between the two systems

- Byte-stream support

— End-to-end flow control

— Congestion control

— Transmission of data is a continuous stream of bytes and not independent packets

— Maintenance of transmitted byte sequence is handled in the Sequence Number and Acknowledgment Number fields in the TCP segment header

- Transmission

— Ensures proper delivery of data received from IP to the correct application

— Delivery location is identified by:

- 16-bit number called the *port* or *service number*

- *Source* and *destination ports* are contained in the first word of the segment header

— Protocols most commonly used with specific TCP-defined applications include:

- File Transfer Protocol (FTP)

- Remote Terminal Access (Telnet)

- Simple Mail Transfer Protocol (SMT)

- User Datagram Protocol (UDP)

- Used in more reliable circuits
- Used only the services in the lower layer of the OSI model

Correctly passing data to and from the Application Layer is an important part of what the Transport Layer services do.

Internet Protocol (IP) Functions

- Format and addressing used in Network Layer 3 packets for moving data within the network
- Usually implemented within datagram networks
- Address structure
  - 32 bits
  - Divided into two fields:
    - Network field
      - Identifies the network connected to the Internet
    - Host field
      - Identifies a particular host belonging to that network. It is subdivided into:
        - Subnet
          - Set to 1
          - Remaining bits are set to 0
        - Subnet address masking
          - Corresponds to the network
          - Significant only to local network provider
          - Requires network devices to be properly optioned to recognize the mask

TCP/IP 7-Layer Responsibilities

No direct correlation exists between the TCP protocol suite and the OSI protocol suite. However OSI and the TCP/IP responsibilities are compared in Figure 17.2.

OSI	TCP/IP 7 Layer Responsibilities
Application	Process
Presentation	
Session	Host to Host Layer
Transport	
Network	IP
Data Link	Network Interface
Physical	

The TCP Process Layer is roughly equivalent to the OSI Presentation and Application Layers. The Host-to-Host Layer performs a function similar to the OSI Transport and Session Layers. The Internet Layer functions (with ARP and ICMP) span the OSI Data Link, Network, and Transport Layers. Parts of the OSI Physical and Data Link Control Layers are in the network interface layer of TCP/IP.

The Process Layer

When referring to the OSI model, the Process Layer in TCP/IP is a combination of:

- Application Layer
- Presentation Layer.

APPLICATION LAYER. The OSI Application Layer is the upper layer of the TCP/IP Process Layer. Application Layer responsibilities are:

- To provide a transparent interface between the user and the network
- To allow the user to designate the task and the application that performs the desired job
- Those major TCP/IP processes and their definitions that are *designated by the user* include:
  - Simple Mail Transfer Protocol (SMTP)
    - Sends electronic mail
  - Trivial File Transfer Protocol (TFTP)
    - Transfers a file
  - HyperText Transfer Protocol (HTTP)
    - Browse the World Wide Web
  - Telnet
    - Terminal access to a remote server and its functions
  - File Transfer Protocol (FTP)
    - Provides for connection-oriented, reliable transfer of files
- Those major TCP/IP processes and their definitions that are *designated by the user, System Administrator, and/or Network Administrator* include:
  - Domain Name Service (DNS)
    - Translates names into IP addresses
  - Boot Protocol (BootP)
    - Offers stations the data they need to join the network

- Simple Network Management Protocol (SNMP),

    - Manages network devices

- Routing Information Protocol (RIP)

    - Allows routers to share routing information

PRESENTATION LAYER. The OSI Presentation Layer is the lower layer of the TCP/IP Process Layer. Presentation Layer responsibilities are:

- To provide common user services to the network user

    - Users services are not essential to the network connection

    - May be useful in a nonnetworking environment

    - Common Presentation Layer services include:

        - Text Compression

            • Compressing a file so that it takes up less room on a disk and requires fewer bits for transmission

        - Character-Code Conversion

            • Translation between ASCII, EBCDIC, Baudot (transcode), and other character codes

        - Encryption

            • Coding and decoding of transmissions

            • Uses private encryption techniques such as:

                • Data Encryption Standard (DES)

                • Public-key techniques

Host-to-Host Layer

There are two OSI Layers referenced in the Host-to-Host Layer of the TCP/IP responsibility model. These two layers are:

- Session Layer
- Transport Layer

SESSION LAYER.   When compared to the OSI model, the TCP/IP Host-to-Host Layer is a combination of the Session Layer and the Presentation Layer. Functions performed in the Session Layer include:

- The connection of an active process within one host to communicate with an active process from another host

- The connection between the two end-communicating processes

- Operation at highest layer that actually deals with a network connection

- No identification as a separate layer in the TCP/IP protocol hierarchy

    — In TCP/IP, this function largely occurs in the Transport Layer, and the term "session" is not used

- Standard data presentation routines that are handled within the TCP/IP applications

- Resynchronization

    — In case of lower layer failure, the Session Layer must know which messages have been received

- Recovery

    — If the Transport Layer connection fails, this layer may have to establish a new transport connection without notifying the higher layers or the user

- Normal and expedited data exchange

    — Must determine whether a given message is to be handled in a normal or expedited manner

    — There must be an agreement on data representation for applications to properly exchange data

    — Manages the sessions (connections) between cooperating applications

    — In TCP/IP, the terms *socket* and *port* are used to describe the path over which cooperating applications communicate

- Authentication

    — Upon establishment of a session, this layer must ensure that the caller has the required privileges to set up a connection

    — Also upon establishment of a session, appropriate resources must be made available

- Billing

    — Upon connection establishment, the appropriate billing algorithm must be employed

        - Example: Billing may be calculated by:

            • Length-of-call basis

            • Per-packet basis

    — Station to be billed must also be identified

In OSI, the Session Layer provides standard data presentation routines, which are handled within the TCP/IP applications.

TRANSPORT LAYER. The Transport Layer is responsible for Host-to-Host Layer functions including:

- Providing error-free communication across the subnetwork and between two host systems

- Providing end-to-end flow control

    — Ensures that the transmitting host does not send more messages than the receiving host can handle

- Providing end-to-end delivery

    — Ensures that transmitted messages are delivered

- Segmentation

    — Most data networks transmit entities with some fixed maximum size, called *packets.* Messages, which may be of any length, must be broken down into packets (fragmented) and reassembled at the receiving side

- Network error recovery

- — The Transport Layer must recover if a failure occurs in the subnetwork
- ▪ Line multiplexing
  - — The transport Layer must determine the optimal use of communication facilities
    - May include several slow processes sharing one high-speed channel
    - May allow one high-speed process to utilize several channels
- ▪ Sequencing
  - — Order of messages sent between hosts must be preserved
- ▪ Process connection
  - — Provides the appropriate end-to-end connection between the end-communicating processes

The Transport Layer is the lowest of the end-to-end layers. Levels 4 through 7 are implemented in host systems only.

The Transport Layer has two protocols: TCP and UDP. Most applications are written to use either of these two protocols. Both these protocols have functions in common:

- ▪ Interface directly with IP, which is where all data flows
- ▪ IP delivers data from the upper layers to the correct network
- ▪ IP also delivers data from the network to the correct transport service
- ▪ Transport services deliver the data they receive from IP to the correct application

Figure 17.3 shows the Process Layer protocols in relation to the Host-to-Host Layer protocols.

Figure 17.3
Process Layer and
Host-to-Host
protocols.

Process Layer	Telnet	FTP	HTML	FTP	DNS	BOOTP	TFTP	RIP
Transport and Session Layer	TCP				UDP			

When determining when to use TCP or UDP, there are two main factors to be considered:

- Speed
  - UDP is a faster protocol than TCP because it does not require a returned receipt for its services
- Reliability
  - TCP is reliable because it accounts for bytes received by sending an acknowledgment

The decision is typically based on the source and target locations. In the same network, reliability is known and so speed can be optimized. In connecting to another network, reliability is less assured and thus becomes more important.

Internet Layer

The Internet Layer compares with the Network Layer in the OSI model. Functions that are performed in the Internet Layer include:

- Connection management across the network and isolation of the upper layer protocols from the details of the underlying network
- The addressing and delivery of data
- TCP header added to the user's message; it is then passed to the Internet Layer for processing

- Internet Layer header added to the TCP segment to create an IP datagram

- Fragmentation

  — Prevents sending additional data to the Network Interface Layer than it can handle in one datagram

  — Usually occurs in routers as they send datagrams from one interface protocol to another

- Addressing

  — Address Resolution Protocol (ARP) assists the Source IP entity to find the physical address that matches the given target IP address

  — Internet Protocol (IP) adds the source and target IP addresses to the IP header routing

- Routing

  — Routers read address provided by the Internet Protocol (IP) so that they know where to send the datagram packets

Figure 17.4 shows the relationship of Host-to-Host and Internet Layer protocols.

Figure 17.4
Host-to-Host Layer and Internet Protocol Layer protocols.

Host-to-Host Layer	TCP		UDP			Segment
Internet Protocol Layer	ARP	OSPF	EGP	ARP	ICMP on IP Layer only	Datagram

Network Interface Layer

The Network Interface Layer is a combination of:

- Data Link Control Layer

- Physical Layer

DATA LINK CONTROL LAYER.   The OSI Data Link Control Layer is located in the upper TCP/IP Network Layer. Some of the Data Link Control Layer's responsibilities are:

- Responsibility for handling the reliable delivery of data across the underlying physical network

- TCP/IP rarely creates protocols in the Data Link Layer—Request for Comments (RFCs) that relate to the Data Link Layer make use of existing data link protocols

PHYSICAL LAYER.   The OSI Physical Layer is located in the lower TCP/IP Network Layer. Some of the Physical Layer's responsibilities are:

- Definition of the characteristics of the hardware needed to carry the data transmission signal

  — Voltage levels and the number and location of interface pins are defined in this layer

  — Standards interface connectors such as RS232C and V.35 and standard local area network wiring, such as IEEE 802.3

  — TCP/IP does not define physical standards—it makes use of existing standards

- Receives the datagram packet from the Internet Protocol (IP) and places it on the network

- Verification of the packet's target hardware address

- Verification of the Internet Protocol (IP) checksum

- Reassembly of fragments

- Verification of the target Internet Protocol (IP) address

- Verification of the Host-to-Host layer protocol

- Transmission of datagram to the identified protocol in the Internet Layer

Figure 17.5 illustrates the relationship of protocols in the Internet Protocol Layer and the Network Interface Layer.

Figure 17.5
Internet Protocol
Layer and Network
Interface Layer
protocols.

Internet Ptotocol Layer	TCP	UDP	RARP	Datagram
Network Interface Layer	Ethernet, Token Ring, Bus, etc.			Packet

Data to Network Flow

Figure 17.6 illustrates the data to network flow in TCP/IP.

Figure 17.6
Flow of data through
a TCP/IP network.

Data → Messages → Segments → Datagrams → Packets → Out Onto Network

The data flow begins with the Process Layer application on the source (client) talking with the Process Layer application on the target (server)—this logic continues down the two stacks. The matching entities on the two hosts talk to each other using the layers below them to carry the conversation. In other words, UDP on one host does not talk with TCP on the other.

Variable Lengths in TCP/IP Headers

Data and application header lengths may vary because:

- Length is limited by the *Maximum Transmission Unit* (MTU) size of the Network Interface Layer

- Internet Layer offers fragmentation to accommodate the size requirements of the MTU

- Other header sizes are more predictable

- UDP header may be in the packet instead of the TCP header

- UDP header is always 8 bytes long

- IP header is 20 bytes by default and only expands to allow for options

Figure 17.7 shows how the TCP/IP header may vary in size.

Figure 17.7
The TCP/IP header may vary in size.

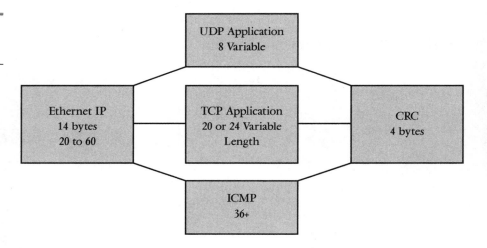

IP Addressing

To deliver data between two Internet hosts, it is necessary to move the data across the network to the correct host, and within that host to the correct user or process. TCP/IP uses three schemes to accomplish these tasks:

- Addressing

  — IP addresses deliver data to the correct host

- Routing

  — Gateways deliver data to the correct network

- Multiplexing

  — Protocol and port numbers deliver data to the correct software module within the host

The standard structure of an IP address can be locally modified by using host address bits as additional network address bits. When network address bits and host address bits are moved, they create additional networks while reducing the maximum number of hosts that can belong to each new network. These newly designated network bits define a network within the larger network, called a *subnet*.

IP addresses are written in byte-based decimal (binary) format (see Figure 17.8). IP addresses are transmitted by using four bytes (32 bits) of data. The first byte designates the class of an IP address. This information is useful in determining which bits are network bits and which bits are locally administered. There are four classes in active use today, so these determinations are necessary.

Local bits can identify interfaces on a network. Some network administrators use part of the local bits to create more manageable subnets: There may be network, subnet, and host fields in an IP address. Some rules apply to the IP address and those fields:

- No field of an interface's IP address may contain all 1s or all 0s (binary)

- All 1s in the host portion of a target IP address signify an IP-level broadcast

- All 0s in the first portion of an IP address identify a subnet or a network

Figure 17.8
Binary format of an IP
address.

Decimal	126	136	118	123
Binary	01111110	10001000	01110110	01111011
Logical	Network	Subnet	Subnet	Interface

IP Address

Internet Protocol (IP) addresses are assigned by the Network Information Center (NIC). There are four classes of Internet address: Class A through Class D. These class address assignments are based on the first numbers of a Internet Protocol (IP) address.

CLASS A NETWORKS AND ADDRESSES

- Identified by 0 as the first bit of an IP address
- Also identified by a first byte value of 1 through 127
- Number 127 Class A network is reserved for IP loopback testing
- The first byte (eight bits) is assigned and three bytes (24 bits) are locally administered
- Using only host addresses, there could be 16,777,214 host addresses (with all 0s and all 1s eliminated)
- Only assigned to large organizations, with many subnets or hosts
- To manage a large network, the network administrator frequently separates the network into smaller subnets

CLASS B NETWORKS AND ADDRESSES

- Identified by 1 as the first bit and 0 as the second bit
- The first byte is in a range of decimal values of 128 through 191
- Assignments include the first two bytes, for 16,384 possible Class B addresses with two bytes of local space each

Figure 17.9
Class A IP address format.

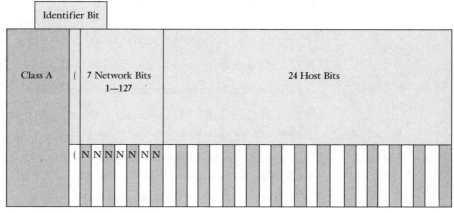

N = Network; L = Locally Administered.

- Without subnetting, this equals 65,534 addresses on a flat network
- In a network this large, the network is separated into subnets
- Subnets help control workgroup access to certain resources
- Assigned to midsize organizations such as colleges and universities having a modest number of subnets or hosts

Figure 17.10
Class B IP address format.

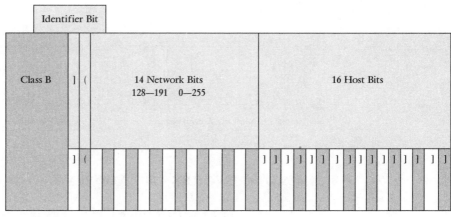

N = Network; L = Locally Administered.

CLASS C NETWORKS AND ADDRESSES

- First two bits are 1s and the third bit is 0

- First byte is in a range of decimal values 192 through 223

- Class C address covers the first three bytes

- There are 2,097,152 possible Class C addresses with one byte of locally administered address space each

- One byte can provide 254 host addresses for each Class C network

- Class C networks do not need subnetting for management unless there are smaller workgroups in diverse locations

- Organizations usually subnet Class C networks to restrict access to specific resources

- Assigned to clients with a small number of subnets and/or hosts

Figure 17.11
Class C IP address format.

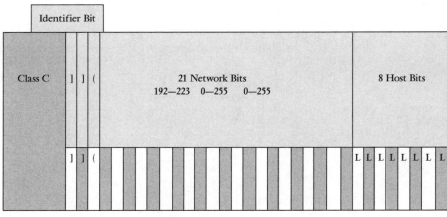

N = Network; L = Locally Administered.

CLASS D NETWORKS AND ADDRESSES

- Identified by the first three bits set to 1 and the fourth bit set to 0

- The first byte is in a range of decimal values 224 through 239

- Used to reach groups by assigning the same multicast address to all members of the group

- These group members also have their own individual Class A, B, or C host IP address

- There are millions of possible multicast addresses

- Class D addresses are designated for groups of users

- Class D addresses do not have assignable host portions for individual interfaces

- Class D networks are not subnetted

- Class D defines a specific host

Figure 17.12
Class D IP address format.

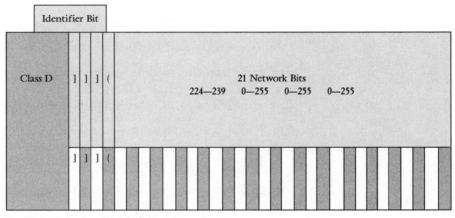

N = Network; L = Locally Administered.

Subnetting

An IP addressed interface that wants to communicate with any other IP addressed interface must follow this process.

ANALYZE TARGET IP ADDRESS

- Originating software checks that it is in the target's IP class to determine which bits are network bits in that address

- Originating software determines if the source and target IP addresses are in the same network

  — If not in the same network, the IP datagram is sent to the gateway for delivery to the right network

  — If in the same network, the system must decide if the IP addresses are in the same subnet

ANALYZE SUBNETTING POSSIBILITIES

- Situations that classify the need to subnet

  — Geographically remote from each other

  — Functional areas requiring separation

  — High-level traffic

  — Multiple media protocols connected with each other

  — Connection of multiple segments

- To recognize the subnet, each system in the subnet must have the same *subnet mask* which identifies which bits are network bits, subnet bits, and interface bits (see Figure 17.13)

- Mask must carry the same number of bits as the IP address (32) and is usually named in the same format as the IP address

- The mask's first byte always has a value of 255 (decimal) or FF (hex), which is not possible for an IP addressed interface

- In the structure of a mask

  — Binary 1s indicate the position of the network and subnet portion of the IP address

  — Binary 0s identify bits that represent individual interfaces

To determine proper subnetting for a network, the Network Administrator must address the following issues:

- The limits of the network class that is to be subnetted and the rules for IP addressing

- The quantity of subnets the organization needs from this network

Figure 17.13
Mask structure.

Mask Decimal	255	255	255	0
Mask Hex	ff	FF	FC	00
Mask Binary	11111111	11111111	11111100	00000000
Mask Meaning	NNNNNNNN	NNNNNNNN	SSSSSSII	IIIIIIII
IP Binary	10111111	11111111	11000001	00101100
IP Decimal	191	255	193	44

N = Net; S = Subnet; I = Interface.

- The maximum number of interfaces that must be in the largest subnet

To calculate the number of hosts or subnetworks in any IP addressed network, apply the formula below:

$$(2^n) - 2 = \text{the number of subnets or hosts in a subnet}$$

$$n = \text{the number of bits used in the mask}$$

Masking

The Network Administrator uses a mask to assign IP addresses to individual interfaces. A good guide is to begin identifying subnets from the highest order bits (left) and interfaces from the lowest order bits (right). This method offers the greatest flexibility to adjust mask bits right up to the last available interface assignments. However, the total number of interfaces under subnets is always less than the number of interfaces without subnetting, which means there is a price to be paid for the ability to manage a network more easily.

Routing questions are simplified with the use of a subnet mask. The source and target IP addresses can quickly determine if the two are in the same subnet. If the subnets are different, the IP datagram is sent to the router serving the source system's subnet for forwarding to the correct remote subnet's router and to the target IP address.

Tables 17.1, 17.2, and 17.3 are subnetting charts for Class A, B, and C networks, respectively.

TABLE 17.1   Class A Subnetting Table

Subnet Bits	Subnet Mask	Subnets	Hosts	Subnet Broadcast
2	255.192.0.0	2	4,194,302	net.subnet+63.255.255
3	255.224.0.0	6	2,097,150	net.subnet+31.225.255
4	255.240.0.0	14	1,048.574	net.subnet+15.255.255
5	255.248.0.0	30	524,286	net.subnet+7.255.255
6	255.242.0.0	62	262,142	net.subnet+3.255.255
7	255.254.0.0	126	131,070	net.subnet+1.255.255
8	255.255.0.0	254	65,534	net.subnet+255.255
9	255.255.128.0	510	32,766	net.subnet+127.255
10	255.255.192.0	1,022	16,382	net.subnet+63.255
11	255.255.224.0	2,046	8,190	net.subnet+31.255
12	255.255.240.0	4,094	4,094	net.subnet+15.255
13	255.255.248.0	8,190	2,046	net.subnet+7.255
14	255.255.252.0	16,382	1,022	net.subnet+3.255
15	255.255.254.0	32,766	510	net.subnet+1.255
16	255.255.255.0	65,534	254	net.subnet.255
17	255.255.255.128	132,070	126	net.subnet+127
18	255.255.255.192	262,142	62	net.subnet+63
19	255.255.255.224	524,286	30	net.subnet+31
20	255.255.255.240	1,048,574	14	net.subnet+15
21	255.255.255.248	2,097,150	6	net.subnet+7
22	255.255.255.252	4,194,302	2	net.subnet+3

TABLE 17.2 Class B Subnetting Table

Subnet Bits	Subnet Mask	Subnets	Hosts	Subnet Broadcast
2	255.255.192.0	2	16,382	net.net.subnet+63.255
3	255.255.224.0	6	8,190	net.net.subnet+31.255
4	255.255.240.0	14	4,094	net.net.subnet+15.255
5	255.255.248.0	30	2,046	net.net.subnet+7,255
6	255.255.252.0	62	1,022	net.net.subnet+3.255
7	255.255.254.0	126	510	net.net.subnet+1.255
8	255.255.255.0	254	254	net.net.subnet.255
9	255.255.255.128	510	126	net.net.subnet+127
10	255.255.255.192	1,022	62	net.net.subnet+63
11	255.255.255.224	2,046	30	net.net.subnet+31
12	255.255.255.240	4,094	14	net.net.subnet+15
13	255.255.255.248	8,190	6	net.net.subnet+7
14	255.255.255.252	16,382	2	net.net.subnet+3

TABLE 17.3 Class C Subnetting Table

Subnet Bits	Subnet Mask	Subnets	Hosts	Subnet Broadcast
2	255.255.255.192	2	62	net.net.net.subnet+63
3	255.255.255,224	6	30	net.net.net.subnet+31
4	255.255.255.240	14	14	net.net.net.subnet+15
5	255.255.255.248	30	6	net.net.net.subnet+7
6	255.255.255.252	62	2	net.net.net.subnet+3

Protocols, Ports, and Sockets

To complete data transmission to the correct user or process on a host the following mechanisms are required:

- *Protocol numbers* are preassigned to identify transport protocols that move data up or down the layers of TCP/IP

- *Port numbers* are preassigned to identify data applications and transport protocols into the Internet

- *Dynamically allocated port numbers* are not preassigned. Port number assignment is made when needed. The system ensures that it does not assign the same port number to two processes, and that the numbers assigned are above the range of standard port numbers (1—1023)

- *Multiplexing* is used to combine many sources of data into a single data stream

- *Demultiplexing* divides data arriving from the network for delivery to multiple processes

Standard protocol numbers and port numbers that are allocated to common services are documented in the Assigned Numbers RFC. UNIX systems define protocol and port numbers in simple text files. Standardized (or well-known) port numbers enable remote computers to know which port to connect to for a particular network service.

Protocol numbers:

- Occupy a single byte in the third word of the datagram header

- Identify the protocol in the layer above IP to which the data should be passed

- Allow IP to pass incoming data to the transport protocol

- Allow transport protocol to pass the data to the correct application process

Port numbers:

- Are 16-bit values

- Source port number identifies the process that transmitted the data

- Destination port number identifies the process that is to receive the data contained in the first header word of each TCP segment and UDP packet

- Port numbering schemes:

    — 1—256 are reserved for standard services

    — 256—1024 are used for UNIX specific services

A combination of protocol and port numbers identifies the specific process to which the data should be delivered. Table 17.4 lists standard port numbers.

TABLE 17.4

Port Numbers

Network Services		Host Services		UNIX Specific	
ftp-data	20/tcp	Tftp	69/udp	Exec	512/tcp
Smtp	25/tcp	Uucp	117/tcp	Login	513/tcp
Domain	53/udp	Ntp	123/tcp	Route	520/udp

Table 17.5 lists port numbers 1 through 1023. If additional ports are needed, refer to RFC 1700.

TABLE 17.5 Port Assignments

Keywords	Decimal		Description—Reference
	0/tcp	0/udp	Reserved
tcpmux	1/tcp	1/udp	TCP Port Service—Multiplexer
compressnet	2/tcp	2/udp	Management Utility
compressnet	3/tcp	3/udp	Compression Process
rje	5/tcp	5/udp	Remote Job Entry
echo	7/tcp	7/udp	Echo
discard	9/tcp	9/udp	Discard
systat	11/tcp	11/udp	Active Users
daytime	13/tcp	13/udp	Daytime
qotd	17/tcp	17/udp	Quote of the Day
msp	18/tcp	18/udp	Message Send Protocol
chargen	19/tcp	19/udp	Character Generator
ftp-data	20/tcp	20/udp	File Transfer (default data)
ftp	21/tcp	21/udp	File Transfer (Control)
telnet	23/tcp	23/udp	Telnet
	24/tcp	24/udp	Any private mail system
smtp	25/tcp	25/udp	Simple Mail Transfer
nsw-fe	27/tcp	27/udp	NSW User System FE
msg-icp	29/tcp	29/udp	MSG ICP
msg-auth	31/tcp	31/udp	MSG Authentication
dsp	33/tcp	33/udp	Display Support Protocol
	35/tcp	35/udp	Any Private Printer Server
time	37/tcp	37/udp	Time

continued on next page

TABLE 17.5 Port Assignments (Continued)

Keywords	Decimal		Description—Reference
rap	38/tcp	38/udp	Route Access Protocol
rlp	39/tcp	39/udp	Resource Location Protocol
graphics	41/tcp	41/udp	Graphics
nameserver	42/tcp	42/udp	Host Name Server
nicname	43/tcp	43/udp	Who Is
mpm-flags	44/tcp	44/udp	MPM FLAGS Protocol
mpm	45/tcp	45/udp	Message Processing Module (recv)
mpm-snd	46/tcp	46/udp	MPM (default send)
ni-ftp	47/tcp	47/udp	NI FTP
audited	48/tcp	48/udp	Digital Audit Daemon
Login	49/tcp	49/udp	Login Host Protocol
re-mail-ck	50/tcp	50/udp	Remote Mail Checking Protocol
la-maint	51/tcp	51/udp	IMP Logical Address Maintenance
xns-time	52/tcp	52/udp	XNS Time Protocol
domain	53/tcp	53/udp	Domain Name Server
xns-ch	54/tcp	54/udp	XNS Clearinghouse
isi-gl	55/tcp	55/udp	ISI Graphics Language
xns-auth	56/tcp	56/udp	XNS Authentication
	57/tcp	57/udp	Private Terminal Access
xns-mail	58/tcp	58/udp	XNS Mail
	59/tcp	59/udp	Private File Service
ni-mail	61/tcp	61/udp	NI Mail
acas	62/tcp	62/udp	ACA Services

continued on next page

TABLE 17.5 Port Assignments (Continued)

Keywords	Decimal		Description—Reference
covia	64/tcp	64/udp	Communications Integrator (CI)
tacacs-ds	65/tcp	65/udp	TACACS-Database Service
sql*net	66/tcp	66/udp	Oracle SQL*NET
bootps	67/tcp	67/udp	Bootstrap Protocol Server
Bootpc	68/tcp	68/udp	Bootstrap Protocol Client
tftp	69/tcp	69/udp	Trivial File Transfer
gopher	70/tcp	70/udp	Gopher
netrjs-1	71/tcp	71/udp	Remote Job Service
netrjs-2	72/tcp	72/udp	Remote Job Service
netrjs-3	73/tcp	73/udp	Remote Job Service
netrjs-4	74/tcp	74/udp	Remote Job Service
	75/tcp	75/udp	Any Private Dial Out Service
deos	76/tcp	76/udp	Distributed External Object Store
	77/tcp	77/udp	Any Private RJE Service
vettcp	78/tcp	78/udp	
finger	79/tcp	79/udp	Finger
www-http	80/tcp	80/udp	World Wide Web HTTP
hsts2-ns	81/tcp	81/udp	HOSTS2 Name Server
xfer	82/tcp	82/udp	XFER Utility
mit-ml-dev	83/tcp	83/udp	MIT ML Device
ctf	84/tcp	84/udp	Common Trace Facility
mit-ml-dev	85/tcp	85/udp	MIT ML Device
mfcobol	86/tcp	86/udp	Micro Focus Cobol

continued on next page

TABLE 17.5   Port Assignments (Continued)

Keywords	Decimal		Description—Reference
	87/tcp	87/udp	Any Private Terminal Link
kerberos	88/tcp	88/udp	Kerberos
su-mit-tg	89/tcp	89/udp	SU/MIT Telnet Gateway
dnsix	90/tcp	90/udp	DNSIX Securit Attribute Token Map
mit-dov	91/tcp	91/udp	MIT Dover Spooler
npp	92/tcp	92/udp	Network Printing Protocol
dcp	93/tcp	93/udp	Device Control Protocol
objcall	94/tcp	94/udp	Tivoli Object Dispatcher
supdup	95/tcp	95/udp	SUPDUP
dixie	96/tcp	96/udp	DIXIE Protocol Specification
swift-rvf	97/tcp	97/udp	Swift Remote Virtual File Protocol
tacnews	98/tcp	98/udp	TAC News
metagram	99/tcp	99/udp	Metagram Relay
newacct	100/tcp		(unauthorized use)
hostname	101/tcp	101/udp	NIC Host Name Server
iso-tsap	102/tcp	102/udp	ISO-TSAP
gppitnp	103/tcp	103/udp	Genesis Point-to-Point Trans Net
acr-nema	104/tcp	104/udp	ACR-NEMA Digital Imag. & Comm.
csnet-ns	105/tcp	105/udp	Mailbox Name Name Server
3com-tsmux	106/tcp	106/udp	3COM-TSMUX
telnet	107/tcp	107/udp	Remote Telenet Service
snagas	108/tcp	108/udp	SNA Gateway Access Server
pop2	109/tcp	109/udp	Post Office Protocol — Version 2

continued on next page

TABLE 17.5 Port Assignments (Continued)

Keywords	Decimal		Description—Reference
pop3	110/tcp	110/udp	Post Office Protocol — Version 3
sunrpc	111/tcp	111/udp	SUN Remote Procedure Call
mcidas	112/tcp	112/udp	McIDAS Data Transmission Protocol
auth	113/tcp	113/udp	Authentication Service
audionews	114/tcp	114/udp	Audio News Multicast
sftp	115/tcp	115/udp	Simple File Transfer Protocol
ansanotify	116/tcp	116/udp	ANSA REX Notify
uucp-path	117/tcp	117/udp	UUCP Path Service
sqlserv	118/tcp	118/udp	SQL Services
nntp	119/tcp	119/udp	Network News Transfer Protocol
cfdptkt	120/tcp	120/udp	CFDPTKT
erpc	121/tcp	121/udp	Encore Expedited Remote Pro.Call
smakynet	122/tcp	122/udp	SMAKYNET
ntp	123/tcp	123/udp	Network Time Protocol
ansatrader	124/tcp	124/udp	ANSA REX Trader
locus-map	125/tcp	125/udp	Locus PC-Interface Net Map Ser
unitary	126/tcp	126/udp	Unisys Unitary Login
locus-con	127/tcp	127/udp	Locus PC-Interface Conn Server
gss-xlicen	128/tcp	128/udp	GSS X License Verification
wdgen	129/tcp	129/udp	Password Generator Protocol
cisco-fna	130/tcp	130/udp	Cisco FNATIVE
cisco-tna	131/tcp	131/udp	Cisco TNATIVE
cisco-sys	132/tcp	132/udp	Cisco SYSMAINT

continued on next page

TABLE 17.5 Port Assignments (Continued)

Keywords	Decimal		Description—Reference
statsrv	133/tcp	133/udp	Statistics Service
ingres-net	134/tcp	134/udp	INGRES-NET Service
loc-srv	135/tcp	135/udp	Location Service
profile	136/tcp	136/udp	PROFILE Naming System
netbios-ns	137/tcp	137/udp	NETBIOS Name Service
netbios-dgm	138/tcp	138/udp	NETBIOS Datagram Service
netbios-ssn	139/tcp	139/udp	NETBIOS Session Service
emfis-data	140/tcp	140/udp	EMFIS Data Service
emfis-cntl	141/tcp	141/udp	EMFIS Control Service
bl-idm	142/tcp	142/udp	Britton-Lee IDM
imap2	143/tcp	143/udp	Interim Mail Access Protocol V2
news	144/tcp	144/udp	News
uaac	145/tcp	145/udp	UAAC Protocol
iso-tp0	146/tcp	146/udp	ISO-IP0
iso-ip	147/tcp	147/udp	ISO-IP
cronus	148/tcp	148/udp	CRONUS-SUPPORT
aed-512	149/tcp	149/udp	AED 512 Emulation Service
sql-net	150/tcp	150/udp	SQL-NET
hems	151/tcp	151/udp	HEMS
bftp	152/tcp	152/udp	Background File Transfer Program
sgmp	153/tcp	153/udp	SGMP
netsc-prod	154/tcp	154/udp	NETSC
netsc-dev	155/tcp	155/udp	NETSC

continued on next page

TABLE 17.5 Port Assignments (Continued)

Keywords	Decimal		Description—Reference
sqlsrv	156/tcp	156/udp	SQL Service
knet-cmp		157/udp	KNET/VM Command/Message Protocol
pcmail-srv	158/tcp	158/udp	PCMail Server
nss-routing	159/tcp	159/udp	NSS-Routing
sgmp-traps	160/tcp	160/udp	SGMP-TRAPS
snmp	161/tcp	161/tcp	SNMP
snmptrap	162/tcp	162/udp	SNMPTRAP
cmip-man	163/tcp	163/udp	CMIP/TCP Manager
cmip-agent	164/tcp	164/udp	CMIP/TCP Agent
xns-courier	165/tcp	165/udp	Xerox
s-net	166/tcp	166/udp	Sirius Systems
namp	167/tcp	167/udp	NAMP
rsvd	168/tcp	168/udp	RSVD
send	169/tcp	169/udp	SEND
print-srv	170/tcp	170/udp	Network PostScript
multiplex	171/tcp	171/udp	Network Innovations Multiplex
cl/1	172/tcp	172/udp	Network Innovations CL/1
xyplex-mux	173/tcp	173/udp	Xyplex
mailq	174/tcp	174/udp	MAILQ
vmnet	175/tcp	175/udp	VMNET
genrad-mux	176/tcp	176/udp	GENRAD-MUX
xdmcp	177/tcp	177/udp	X Display Manager Control Protocol
nextstep	178/tcp	178/udp	NextStep Window Server

continued on next page

TABLE 17.5   Port Assignments (Continued)

Keywords	Decimal		Description—Reference
bgp	179/tcp	179/udp	Border Gateway Protocol
ris	180/tcp	180/udp	Intergraph
unify	181/tcp	181/udp	Unify
audit	182/tcp	182/udp	Unisys Audit SITP
ocbinder	183/tcp	183/udp	OCBinder
ocserver	184/tcp	184/udp	OCServer
remote-kis	185/tcp	185/udp	Remote-KIS
kis	186/tcp	186/udp	KIS Protocol
aci	187/tcp	187/udp	Application Communication Interface
mumps	188/tcp	188/udp	Plus Five's MUMPS
qft	189/tcp	189/udp	Queued File Transport
gacp	190/tcp	190/udp	Gateway Access Control Protocol
prospero	191/tcp	191/udp	Prospero Directory Service
osu-nms	192/tcp	192/udp	OSU Network Monitoring System
srmp	193/tcp	193/udp	Spider Remote Monitoring Protocol
irc	194/tcp	194/udp	Internet Relay Chat Protocol
dn6-nlm-aud	195/tcp	195/udp	DNSIX Network Level Module Audit
dn6-smm-red	196/tcp	196/udp	DNSIX Session Mgmt Module Audit Redir
Dls	197/tcp	197/udp	Directory Location Service
dls-mon	198/tcp	198/udp	Directory Location Service Monitor
smux	199/tcp	199/udp	SMUX
src	200/tcp	200/udp	IBM System Resource Controller
at-rtmp	201/tcp	201/udp	AppleTalk Routing Maintenance

continued on next page

TABLE 17.5   Port Assignments (Continued)

Keywords	Decimal		Description—Reference
at-nbp	202/tcp	202/udp	AppleTalk Name Binding
at-3	203/tcp	203/udp	AppleTalk Unused
at-echo	204/tcp	204/udp	AppleTalk Echo
at-5	205/tcp	205/udp	AppleTalk Unused
at-zis	206/tcp	206/udp	AppleTalk Zone Information
at-7	207/tcp	207/udp	AppleTalk Unused
at-8	208/tcp	208/udp	AppleTalk Unused
tam	209/tcp	209/udp	Trivial Authenticated Mail Protocol
z39.50	210/tcp	210/udp	ANSI z39.50
914c/g	211/tcp	211/udp	Texas Instruments 914C/G Terminal
anet	212/tcp	212/udp	ATEXSSTR
ipx	213/tcp	213/udp	IPX
vmpwscs	214/tcp	214/udp	VM PWSCS
softpc	215/tcp	215/ucp	Insignia Solutions
atls	216/tcp	216/udp	Access Technology License Server
dbase	217/tcp	217/udp	dBASE Unix
mpp	218/tcp	218/udp	Netix Message Posting Protocol
uarps	219/tcp	219/udp	Unisys ARPs
imap3	220/tcp	220/udp	Interactive Mail Access Protocol V3
fln-spx	221/tcp	221/udp	Berkeley rlogind with SPX auth
rsh-spx	222/tcp	222/udp	Berkeley rshd with SPX auth
cdc	223/tcp	223/udp	Certificate Distribution Center
	224-241/tcp	224-241/udp	Reserved

continued on next page

TABLE 17.5 Port Assignments (Continued)

Keywords	Decimal		Description—Reference
sur-meas	243/tcp	243/udp	Survey Measurement
link	245/tcp	245/udp	LINK
dsp3270	246/tcp	246/udp	Display Systems Protocol
	247-255/tcp	247-255/udp	Reserved
pdap	344/tcp	344/udp	Prospero Data Access Protocol
pawserv	345/tcp	345/udp	Perf Analysis Workbench
zserv	346/tcp	346/udp	Zebra server
fatserv	347/tcp	347/udp	Fatmen Server
csi-sgwp	348/tcp	348/udp	Cabletron Management Protocol
clearcase	371/tcp	371/udp	Clearcase
ulistserv	372/tcp	372/udp	UNIX Listserv
legent-1	373/tcp	373/udp	Legent Corporation
legent-2	374/tcp	374/udp	Legent Corporation
hassle	375/tcp	375/udp	Hassle
nip	376/tcp	376/udp	Amiga Envoy Network Inquiry Protocol
tnETOS	377/tcp	377/udp	NEC Corporation
dsETOS	378/tcp	378/udp	NEC Corporation
is99c	379/tcp	379/udp	TIA/EIA/IS-99 Modem Client
is99s	380/tcp	380/udp	TIA/EIA/IS-99 Modem Server
hp-collector	381/tcp	381/udp	HP Performance Data Collector
hp-managed-node	382/tcp	382/udp	HP Performance Data Managed Node
hp-alarm-mgr	383/tcp	383/udp	HP Performance Data Alarm Manager

continued on next page

TABLE 17.5 Port Assignments (Continued)

Keywords	Decimal		Description—Reference
arns	384/tcp	384/udp	A Remote Network Server System
ibm-app	385/tcp	385/tcp	IBM Application
asa	386/tcp	386/udp	ASA Message Router Object Def.
aurp	387/tcp	387/udp	AppleTalk Update-Based Routing Protocol
unidata-ldm	388/tcp	388/udp	Unidata LDM Version 4
ldap	389/tcp	389/udp	Lightweight Directory Access Protocol
uis	390/tcp	390/udp	UIS
synotics-relay	391/tcp	391/udp	SynOptics SNMP Relay Port
synotics-broker	392/tcp	392/udp	SynOptics Port Broker Port
dis	393/tcp	393/udp	Data Interpretation System
embl-ndt	394/tcp	394/udp	EMBL Nucleic Data Transfer
netcp	395/tcp	395/udp	NETscout Control Protocol
netware-ip	396/tcp	396/udp	Novell Netware over IP
mptn	397/tcp	397/udp	Multi Protocol Transmission Network
kryptolan	398/tcp	398/udp	Kryptolan
work-sol	400/tcp	400/udp	Workstation Solutions
ups	401/tcp	401/udp	Uninterruptible Power Supply
genie	402/tcp	402/udp	Genie Protocol
decap	403/tcp	403/udp	decap
nced	404/tcp	404/udp	nced
ncld	405/tcp	405/udp	ncld
imsp	406/tcp	406/udp	Interactive Mail Support Protocol
timbuktu	407/tcp	407/udp	Timbuktu

continued on next page

TABLE 17.5 Port Assignments (Continued)

Keywords	Decimal		Description—Reference
prm-sm	408/tcp	408/udp	Prospero Resource Manager System Man.
prm-nm	409/tcp	409/udp	Prospero Resource Manager Node Man.
decladebug	410/tcp	410/udp	DECLadebug Remote Debug Protocol
rmt	411/tcp	411/udp	Remote MT Protocol
synoptics-trap	412/tcp	412/udp	Trap Convention Port
smsp	413/tcp	413/udp	SMSP
infoseek	414/tcp	414/udp	InfoSeek
bnet	415/tcp	415/udp	BNet
silverplatter	416/tcp	416/udp	Silverplatter
onmux	417/tcp	417/udp	Onmux
hyper-g	418/tcp	418/udp	Hyper-G
ariel1	419/tcp	419/udp	Ariel
smpte	420/tcp	420/udp	SMPTE
ariel2	421/tcp	421/udp	Ariel
ariel3	422/tcp	422/udp	Ariel
opc-job-start	423/tcp	423/udp	IBM Operations Planning & Control Start
opc-job-track	424/tcp	424/udp	IBM Operations Planning & Control Track
icad-el	425/tcp	425/udp	ICAD
smartsdp	426/tcp	426/udp	smartsdp
svrloc	427/tcp	427/udp	Server Location
ocs_cmu	428/tcp	428/udp	OCS_CMU
ocs_amu	429/tcp	429/udp	OCS_AMU
utmpsd	430/tcp	430/udp	UTMPSD

continued on next page

TABLE 17.5 Port Assignments (Continued)

Keywords	Decimal		Description—Reference
utmpcd	431/tcp	431/udp	UTMPCD
iasd	432/tcp	432/udp	IASD
nnsp	433/tcp	433/udp	NNSP
mobileip-agent	434/tcp	434/udp	MobileIP-Agent
mobilip-mn	435/tcp	435/udp	MobilIP-MN
dna-cml	436/tcp	436/udp	DNA-CML
comscm	437/tcp	437/udp	comscm
dsfgw	438/tcp	438/udp	
dasp	439/tcp	439/udp	
sgcp	440/tcp	440/udp	
decvms-sysmgt	441/tcp	441/udp	
cvc_hostd	442/tcp	442/udp	
https	443/tcp	443/udp	
snpp	444/tcp	444/udp	Simple Network Paging Protocol
microsoft-ds	445/tcp	445/udp	Microsoft-DS
ddm-rdb	446/tcp	446/udp	
ddm-dfm	447/tcp	447/udp	
ddm-byte	448/tcp	448/udp	
as-servermap	449/tcp	449/udp	AS Server Mapper
tserver	450/tcp	450/udp	TServer
Exec	512/tcp	512/udp	Notify Users of New Mail
login	513/tcp		Remote Login a la telnet
who		513/udp	Shows Who Is Logged In

continued on next page

TABLE 17.5 Port Assignments (Continued)

Keywords	Decimal		Description—Reference
cmd	514/tcp		Like exec, But Automatic
syslog		514/udp	
printer	515/tcp	515/udp	Spooler
talk	517/tcp	517/udp	Like tenex Link, But Across Machine
ntalk	518/tcp	518/udp	
utime	519/tcp	519/udp	unixtime
efs	520/tcp	520/udp	Local Routing Process (on site)
timed	525/tcp	525/udp	Timeserver
tempo	526/tcp	526/udp	newdate
courier	530/tcp	530/udp	rpc
conference	531/tcp	531/udp	chat
netnews	532/tcp	532/udp	Read News
netwall	533/tcp	533/udp	For Emergency Broadcasts
apertus-ldp	539/tcp	539/udp	Apertus Technologies Load Determination
uucp	540/tcp	540/udp	uucpd
uucp-rlogin	541/tcp	541/udp	
klogin	543/tcp	543/udp	
kshell	544/tcp	544/udp	krcmd
new-rwho	550/tcp	550/udp	
dsf	555/tcp	555/udp	
remotefs	556/tcp	556/udp	rfs server
rmonitor	560/tcp	560/udp	
monitor	561/tcp	561/udp	

continued on next page

TABLE 17.5 Port Assignments (Continued)

Keywords	Decimal		Description—Reference
chshell	562/tcp	562/udp	chcmd
9pfs	564/tcp	564/udp	Plan 9 File Service
whoami	565/tcp	565/udp	
meter	570/tcp	570/udp	demon
meter	571/tcp	571/udp	udemon
ipcserver	600/tcp	600/udp	Sun IPC server
urm	606/tcp	606/udp	Cray Unified Resource Manager
nqs	607/tcp	607/udp	
sift-uft	608/tcp	608/udp	Sender-Initiated/Unsolicited File Transfer
npmp-trap	609/tcp	609/udp	npmp-trap
npmp-local	610/tcp	610/udp	npmp-local
npmp-gui	611/tcp	611/udp	
ginad	634/tcp	634/udp	
mdqs	666/tcp	666/udp	
elcsd	704/tcp	704/udp	Error Log Copy/Server Daemon
entrustmanager	709/tcp	709/udp	Entrust Manager
netviewdm1	729/tcp	729/udp	IBM NetView DM/6000 Server/Client
netviewdm2	730/tcp	730/udp	IBM NetView DM/6000 Send/TCP
netviewdm3	731/tcp	731/udp	IBM NetView DM/6000 Receive/TCP
netgw	741/tcp	741/udp	
netrcs	742/tcp	742/udp	Network Based Rev. Cont. System
flexlm	744/tcp	744/udp	Flexible License Manager
fujitsu-dev	747/tcp	747/udp	Fujitsu Device Control

continued on next page

TABLE 17.5   Port Assignments (Continued)

Keywords	Decimal		Description—Reference
ris-cm	748/tcp	748/udp	Russell Info Sci Calendar Manager
kerberos-adm	749/tcp	749/udp	Kerberos Administration
rfile	750/tcp	750/udp	
pump	751/tcp	751/udp	
qrh	752/tcp	752/udp	
rrh	753/tcp	753/udp	
tell	754/tcp	754/udp	Send
nlogin	758/tcp	758/udp	
con	759/tcp	759/udp	
ns	760/tcp	760/udp	
rxe	761/tcp	761/udp	
quotad	762/tcp	762/udp	
cycleserv	763/tcp	763/udp	
omserv	764/tcp	764/udp	
webster	765/tcp	765/udp	
phonebook	767/tcp	767/udp	Phone
vid	769/tcp	769/udp	
cadlock	770/tcp	770/udp	
rtip	771/tcp	771/udp	
cycleserv2	772/tcp	772/udp	
submit	773/tcp	773/udp	
rpasswd	774/tcp	774/udp	
entomb	775/tcp	775/udp	

continued on next page

TABLE 17.5 Port Assignments (Continued)

Keywords	Decimal		Description—Reference
wpages	776/tcp	776/udp	
wpgs	780/tcp	780/udp	
concert	786/tcp	786/udp	Concert
mdbs_daemon	800/tcp	800/udp	
device	801/tcp	801/udp	
xtreelic	996/tcp	996/udp	Central Point Software
maitrd	997/tcp	997/udp	
busboy	998/tcp	998/udp	
garcon	999/tcp	999/udp	Applix ac
puprouter	999/tcp	999/udp	
cadlock	1000/tcp	1000/udp	
	1023/tcp		Reserved
		1024/udp	Reserved

Network Address Translation (NAT)

For the IP logic process to take place, the user must enter an IP address or a name to be translated so that system has a target interface. There are two different methods to translate a name:

- Source software checks the HOSTS file for a cross reference to the IP address

  - Each operating system vendor has it own method for translating names and its own name for the HOSTS file (see Table 17.6)

155.90.75.192	Wayne
155.90.86.75	Crystal
155.90.93.86	Travis
192.180.176.78	Compu-Connections.com

- *Domain Name System* (DNS) is a distributed database on the network used by TCP/IP applications to convert names into IP addresses

    — Distributed, so that no single Internet site or Domain Name Server knows all the information

    — Each name must start with an alphabetic character and may contain the 26 alphabetic (letters), 10 numeric (0,1,2,3,4,5,6,7,8,9) characters, and a hyphen (-). No other characters are allowed. Upper- or lowercase characters are not differentiated

    — Names are separated by a period or dot

    — Every node must have a unique domain name, although labels may be used more than once in the tree

    — Top-level domains are supported by DS.INTERNIC.NET

    — *Network zones* are delegated to each network for maintenance

        - A zone is a separately administered portion of the domain name server tree

        - Each zone has at least one domain name server

        - Many zones have multiple servers to prevent a single point of failure

Once the IP address has been determined, each layer of the TCP/IP stack passes that address down the stack to the IP layer. The IP layer then begins the logic process that continues after having determined that the source and target IP addresses are in the same network and the same subnet.

When one host wants to communicate with another host on the same segment of network cable, it must follow these rules:

- The host must know the exact address

- IP needs to match the IP address to a physical address

- Physical address is stored in a volatile RAM space known as the *ARP cache*

- ARP cache offers IP virtually instant access to the physical address that it must pass to the Network Interface Layer

Domain Addresses

Organizational and country *domains* are separate organizational categories structured to reduce the time a DNS resolver requires to find the right network address. By grouping similar organizations and countries, the DNS resolver avoids having to search through all possible Internet domains to find a university file server for a Telnet session, for example.

Table 17.7 lists of the most common organizational domains. Table 17.8 lists country domains.

TABLE 17.7

Organizational
Common Domains

Organization	Domain Name
Commericial Organizations	.com
Educational Institutions	.edu
Non-Military Government Agencies	.gov
International Organizations	.int
U.S. Military	.mil
Networks	.net
Non-Profit Organizations	.org
	continued on next page

TABLE 17.7

*Organizational
Common Domains
(Continued)*

Organization	Domain Name
Cultural Entertainment Entities	.arts
Service Businesses	.firm
Information Providers	.info
Individual or Personal Naming	.nom
Recreational Entertainment Entities	.rec
Businesses Selling Goods	.store
WWW entities	.web

TABLE 17.8

Country Domains

Country	Domain Name	Country	Domain Name
Austria	.at	Ireland	.ie
Australia	.au	Israel	.il
Canada	.ca	India	.in
Costa Rica	.cr	Italy	.it
Denmark	.dk	Japan	.jp
Germany	.de	Korea (south)	.kr
Spain	.es	New Zealand	.nz
Finland	.fi	Sweden	.se
France	.fr	Singapore	.sg
Greece	.gr	Taiwan	.tw
Hong Kong	.hk	United Kingdom	.uk
Croatia	.hr	United States	.us

Address Resolution Protocol (ARP)

The Address Resolution Protocol (ARP) stores a recently accessed source IP and its hardware address in its cache.

- This procedure allows quick access for the source address to determine the target IP and hardware address information instead of having to work through the entire address resolution process

- ARP is usually discarded if the request arrives at a system other than the one with the matching IP address

- If the ARP request is received by the host with the matching target IP address, the ARP cache is updated by adding the source hardware and protocol addresses and setting that entry's timer before responding to the ARP request

- An ARP is only successful when an ARP reply is sent to and received by the host originating the ARP request

Network Layer and Internet Protocol Header Information

The main functions and responsibilities of the Network Layer and Internet Protocol (IP) are:

- Error detection
- Options
- Length identification
- Routing instructions and decisions
- Fragmentation
- Loop prevention
- Protocol identification
- Priority for quality of service

- Logical addressing
- Version identification

Internet Protocol (IP) Header

Each of the protocols in the TCP/IP suite uses a series of bytes (known as a header) to perform its required functions. The IP header is no different.

Table 17.9 details the value of bit placement in the Internet Protocol (IP) header example.

TABLE 17.9 Internet Protocol (IP) Header

3 7 000 00000 00 4b ef 19 00 00 ff 01 88 08 d0 99 b0 99 b8 02 b8 04		
Example	**Placement Name**	**Definition of Placement**
3	IP Version	First hex character sets the version of IP that created the header
7	IP Header Length	Second hex character sets the IP header length as a number of 32-bit data words or 4 octets
000	Data Precedence	Informs receiving IP gateways and routers the importance of the data it is carrying.
00000	Service Type Byte/Octet	Carries the 3-bit Precedence of Data field and the four service type bits of Delay, Throughput, Reliability, and Cost
00 4b	Total IP Length	Total Length of the datagram that includes the IP header and all data behind it. This is a 2-byte field.
ef 19	Datagram ID Number	Host specific field carries the unique ID number of each datagram sent by the host. This is a 2-byte field.
		continue on next page

TABLE 17.9   Internet Protocol (IP) Header (Continued)

Example	Placement Name	Definition of Placement
00 00	Fragmentation	If IP must send a data package that is larger than allowed by the Network Access Layer (NAL) it uses, then the data message must be divided into smaller pieces.
ff	Time to Live	Tells how many seconds the datagram can live before it must be delivered or discarded
01	Protocol Field	Carries the ID number of the higher level protocol. This is an 8-bit field
88 08	IP Header Checksum	Provides error checking on the IP header. Does not cover the data that is carried at the end of the header.
c0 99 b7 02	IP Address	IP Source Address
c0 99 b9 04	IP Address	IP Target Address

Table 17.10 lists the available options for the Data Precedence field.

TABLE 17.10

Data Precedence Options

Binary Value	Field Meaning
111	National Network Control
110	Internetwork Control
101	CRTIC/ECP
100	Flash Override
011	Flash
010	Immediate
001	Priority
000	Routine

Table 17.11 lists four options that are located within the Service Type bits.

TABLE 17.11   Service Type Bits Options

Data Precedence field			Service Type Bits				
			Delay	Throughput	Reliability	Cost	Reserved
0	0	0	1	0	0	0	0
0	0	0	0	1	0	0	0
0	0	0	0	0	1	0	0
0	0	0	0	0	0	1	0

Table 17.12 lists the descriptions of the options available in the Service Type bit field.

TABLE 17.12

Service Type Bits

Service Type	Placement Name	Definition of Placement
10000	Delay Bit	Setting this bit to 1 requests a route with the least amount of propagation delay
01000	Throughput Bit	If this bit is set to 1, supporting IP routers will handle the datagram with the highest throughput
00100	Reliability Bit	Informs the application request that the datagram travel with the least chance of lost data
00010	Cost Bit	RFC 1349 added the use of the Cost Bit to allow the data costs of the sending organization the least amount of dollars to use

Ethernet Address Header

The structure of an Ethernet address is:

- Ethernet addresses contains 6 bytes or octets (48 bits)
- First three bytes (pairs of hexadecimal characters) contain the vendor address component of the network interface card (NIC) address
  - A code can be the same on two or more vendors' NICs
  - A few vendors are not careful about using duplicate registered codes
  - Results are unpredictable if cards are installed on the same side of a router
- Last three bytes carry the serial number of that vendor's card
- Protocol analyzers display Ethernet address as 12 hexadecimal characters
- If the first byte of the Ethernet address field becomes the lowest-valued bit, this indicates that the transmission is a multicast and the destination address is shared by multiple recipients; some systems participate in more than one multicast group

Ethernet Target Hardware Address

The destination Target Hardware Address field is listed first in the Ethernet address so that the NICs know which packets to keep as they watch the network. It may contain:

- Unicast
  - A specific target
- Multicast
  - A group of targets with something in common
  - Addresses contain a multicast bit

— Least significant bit of the most significant byte of the vendor address component of the address

- Broadcast

    — All potential targets within a portion of the network

    — Addresses contain all binary 1s

    — Protocol analyzer displays address as hexadecimal Fs (ff ff ff ff ff ff).

Ethernet Source Hardware Address

The Source Hardware Address field identifies the specific hardware card that originated the Ethernet frame:

- Cannot be set to a multicast value

- Cannot be set to a broadcast value

- Protocol Type field acts as a shipping label to identify what function is to receive the contents of this packet at the target end of the transmission

Table 17.13 shows an example of unicast, multicast, and broadcast Ethernet addresses.

TABLE 17.13

Typical Ethernet Addresses

Packet	Target Hardware Address	Source Hardware Address
Unicast	00 00 c0 a0 51 24	00 00 0c 7d 4d 2c
Multicast	01 00 1d 00 00 00	Target Only
Broadcast	ff ff ff ff ff ff	Target Only

Table 17.14 shows an example of the layout for an Ethernet II Header.

TABLE 17.14 Header Layout for Ethernet II

Target Hardware Address	Source Hardware Address	Protocol Type	Data	CRC
6 Bytes	6 Bytes	2 Bytes	46-1500 Bytes	4 Bytes
00 00 e0 b0 41 14	00 00 0d 3c 5d 3d	08 00	IP	01 6b 5e 79

Table 17.15 shows an example of the layout for 802.3 Header.

TABLE 17.15 Typical Ethernet 802.3 Header Layout

Target Hardware Address	Source Hardware Address	Data Length	802.2 LLC	SNAP	DATA	CRC
6 Bytes	6 Bytes	2 Bytes	3 Bytes	5 Bytes	38-1492 Bytes	4 Bytes
00 c0 c0 a0 41 24	00 d0 0c 4a 7d 3c	00 7b	IP	3	2	01 8c 7d 88

Packet Size

Ethernet and IEEE rules set limits on the size of the data packet that may be carried and the overall size of the packet that includes the data:

- Maximum Transmission Unit (MTU) specifies that an Ethernet and 802.3 packet may contain up to 1,500 bytes of data (when 802.3 is not carrying TCP/IP)

- If an Ethernet NIC discovers a packet larger than 1,518 bytes (14-byte header, 1,500 bytes of data, and a 4-byte checksum) the packet is ignored because it is too large

- — IP prevents packets from being too large by fragmenting the data into pieces that fit the MTU

- If an Ethernet NIC discovers a packet smaller than 64 bytes (14-byte header, 46 bytes of data, and a 4-byte checksum) the packet is ignored because it is too small

  - — Ethernet driver adds padding to the data to bring it to at least 46 bytes before sending

INDEX

ABOUT THE AUTHOR

Debbra Wetteroth is currently Web Technology manager for a global telecommunications corporation. Her primary function is to standardize web technology software products for the entire enterprise. Previous IT positions held include Corporate Information Security and Application Testing.

Before entering IT, Debbra developed spent 15 years in Central Office Engineering. She has acquired certifications in Telecom, Internet Security, Firewall Administration and Engineering, UNIX System Administration, RoutingManagement, TCP/IP and Web programming. This book is a compilation of the practices that she has studied in the classroom and tested in the field.